FEEDBACK CONTROL SYSTEMS:
A Fast-Track Guide for Scientists and Engineers

FEEDBACK CONTROL SYSTEMS:
A Fast-Track Guide for Scientists and Engineers

by

ALEX ABRAMOVICI, Ph.D.
JAKE CHAPSKY, P.E.

Jet Propulsion Laboratory, California Institute of Technology

KLUWER ACADEMIC PUBLISHERS
Boston / Dordrecht / London

Distributors for North, Central and South America:
Kluwer Academic Publishers
101 Philip Drive
Assinippi Park
Norwell, Massachusetts 02061 USA
Telephone (781) 871-6600
Fax (781) 681-9045
E-Mail <kluwer@wkap.com>

Distributors for all other countries:
Kluwer Academic Publishers Group
Distribution Centre
Post Office Box 322
3300 AH Dordrecht, THE NETHERLANDS
Telephone 31 78 6392 392
Fax 31 78 6546 474
E-Mail <services@wkap.nl>

 Electronic Services <http://www.wkap.nl>

Library of Congress Cataloging-in-Publication Data

Abramovici, Alex.
  Feedback control systems: a fast-track guide for scientists and engineers / by Alex Abramovici, Jake Chapsky.
    p.cm.
  ISBN 0-7923-7935-7 (alk. paper)
   1. Feedback control systems. I. Chapsky, Jake. II. Title.

TJ216 .A27 2000
629.8'3—dc21

00-057785

Copyright © 2000 by Kluwer Academic Publishers

All rights reserved. No part of this publication may be reproduced, stored in a retrieval system or transmitted in any form or by any means, mechanical, photo-copying, recording, or otherwise, without the prior written permission of the publisher, Kluwer Academic Publishers, 101 Philip Drive, Assinippi Park, Norwell, Massachusetts 02061

Printed on acid-free paper.

Printed in the United States of America

To our parents.

Contents

Summary of Notations — x
List of Examples — x
List of Figures — xi
Preface — xv
Acknowledgments — xxiii

Part I INTRODUCTION TO FEEDBACK CONTROL SYSTEMS

1. CONTROL SYSTEM DIAGRAMS — 3
2. SIGNALS IN THE FEEDBACK LOOP — 9
   2.1 Relationships between Signals in a Loop — 9
   2.2 Simplified FCS Diagrams — 13
   2.3 Actuator Output Range — 15
3. STABILITY — 17
   3.1 Differential Equations, Laplace Transforms and Stability — 17
   3.2 The Nyquist Stability Criterion and Bode Diagrams — 19
   3.3 System Stability: Intuitive Approach — 26
4. EXAMPLES — 35
   4.1 Camera for Aircraft Tracking — 35
       4.1.1 Camera: Range and Lock Acquisition — 35
       4.1.2 Electrical Motors as Actuators — 36
   4.2 Nd:YAG Laser Frequency Stabilization — 39
       4.2.1 Free Running Laser Frequency Noise — 40
       4.2.2 System Concept — 41
       4.2.3 Tracking Requirement — 46

	4.2.4 Environmental Parameters	47
	4.2.5 In-Band, Out-of-Band Frequency Ranges	47

Part II DESIGN AND IMPLEMENTATION

5. DESIGN PRINCIPLES — 51
- 5.1 Design Approach — 51
  - 5.1.1 Assumptions — 52
  - 5.1.2 Error Budget — 53
  - 5.1.3 Design Sequence — 54
- 5.2 Input Data for FCS Design — 57

6. DESIGN AND TROUBLESHOOTING — 59
- 6.1 Sensor Specification — 60
  - 6.1.1 Sensor Range — 60
  - 6.1.2 Sensor Error — 62
  - 6.1.3 Sensor Transfer Function — 64
  - 6.1.4 Sensor Nonlinearity — 66
- 6.2 Actuator Specification — 67
- 6.3 Shaping the Open-Loop Response — 68
- 6.4 Compensator Specification — 72
  - 6.4.1 Compensator Frequency Response Specification — 72
  - 6.4.2 Compensator Hardware Specification — 72
- 6.5 Lock Acquisition — 75
- 6.6 System Integration: Making It All Work — 78
  - 6.6.1 Achieving Closed-Loop Operation — 79
  - 6.6.2 Measuring the Open-Loop Frequency Response — 89
  - 6.6.3 Measuring the Free-Running Variable — 91
  - 6.6.4 Evaluating Tracking Performance — 91
- 6.7 Lock Acquisition Efficiency — 95
- 6.8 Refining the system — 96

7. MULTIPLE SIGNAL PATHS — 99
- 7.1 The Need for Multiple Paths: Examples — 101
  - 7.1.1 Parallel Actuators for Laser Frequency Noise Suppression 101
  - 7.1.2 Parallel Low Frequency Gain Boost — 105
  - 7.1.3 Two Actuators with Nested Loops for Keeping the Optical Path Constant — 108
- 7.2 Parallel and Nested Loops: Equivalence and Stability — 110
  - 7.2.1 Equivalence — 110
  - 7.2.2 Stability Criterion — 113
- 7.3 The PID Compensator — 114
- 7.4 Choosing a Multiple-Path Configuration — 116

Contents   ix

8. DIGITAL COMPENSATORS   119
   8.1 When Should One Use Digital Compensators?   121
   8.2 Aspects of Digital Compensator Design   123
       8.2.1 A/D Converter and Anti-Aliasing Filter   124
       8.2.2 Range Matching Amplifier (RMA)   127
       8.2.3 Digital Filter Block   127
       8.2.4 Digital-to-Analog Converter   132
       8.2.5 Smoothening Filter   132
   8.3 Step-by-Step Specification Recipe   133

Part III   SPECIAL TOPICS

9. ACTIVE NULL MEASUREMENTS   137

10. TWO SENSORS FOR ONE VARIABLE   141
   10.1 Stability of Control Systems with Two Sensors   143
   10.2 Noise Considerations   144

11. FLEXIBLE ELEMENTS AND STABILITY   147
   11.1 Effect of Structure Flexibility on Loop Response   147
   11.2 Resonance-Induced Instability and Ways to Prevent It   150

Appendices   153
   A Poles and Zeros   153
   B Stability of Operational Amplifiers   159
     B.1 OpAmp as Feedback Amplifier   160
     B.2 Unity Gain Stability and Compensation   164
     B.3 Driving Capacitive Loads   165
     B.4 Input Capacitance and Photodiode Preamplifiers   167
   C Quantization Error   171
     C.1 Assumptions and Notations   171
     C.2 Quantization Error Suppression   173
       C.2.1 Signal/Error Ratio with Dithering and Filtering   173
       C.2.2 Signal Recovery   174
     C.3 Example   176

Index   179

Summary of Notations

Variables
$x(t)$: variable, $X(s)$: its Laplace transform
$X(f)$: frequency domain representation of $x(t)$
$e(t), E(s), E(f)$: electronic signals (voltages)
$A(s)$: transfer function, $A(j\omega)$: frequency response
$s = \sigma + j\omega$: Laplace variable
σ: damping, real part of s
ω: $2\pi f$
f: frequency or Fourier frequency
$N(f)$: amplitude spectral density of noise

Subscripts
fr: free-running
Cl: closed-loop
o: output
d: disturbance
r: reference
G: compensator
c: correction
s: sensor

List of Examples

Example	Page
Tracking camera	35
Laser frequency noise suppression	39
Optical path stabilization	108
Two-sensor motion control	142
Control-structure interaction	147
Amplifier with capacitive load	165
Photodiode preamplifier	167

List of Figures

1.1	General schematic representation of a feedback control system.	5
1.2	Algebraic meaning of blocks in feedback control system diagrams.	6
1.3	Example of equivalent blocks.	7
2.1	Schematic representation of a feedback control system, showing the signals and the noise contributions.	11
2.2	Simplified diagrams of reference-following (tracking) and disturbance suppression feedback systems.	14
3.1	Nyquist diagram.	20
3.2	The concepts of phase and gain margins.	20
3.3	Example of Bode plot.	21
3.4	Connection between phase margin and time domain behavior of a closed-loop system.	23
3.5	Example of asymptotic Bode diagram.	25
3.6	Diagram of feedback system used for disturbance suppression.	26
3.7	Forcing function used to examine system stability.	27
3.8	Bode plot of an open-loop transfer function with $0.1°$ phase deficit.	28
3.9	System response to a 4s pulse (0.25 Hz bandwidth).	29
3.10	System response to a 0.5 s pulse (2 Hz bandwidth).	30
3.11	Bode diagram of system with 73 with $0.1°$ phase deficit.	31
3.12	Input signal used to illustrate the effect of a large phase deficit at the unity gain frequency.	32
3.13	Instability build-up for a system with large phase deficit.	33
4.1	Example of a tracking system.	37

xii FEEDBACK CONTROL SYSTEMS

4.2	Azimuth-elevation pointing for CCD camera.	38
4.3	Effect of current/voltage drive for a motor.	39
4.4	Example of upper limit for the free-running frequency noise of a ~ 1 W monolithic Nd:YAG laser.	40
4.5	Concept of laser frequency noise suppression system.	42
4.6	Essential features of the Pound-Drever frequency noise sensor.	43
4.7	Sample Bode plot of output characteristic from a Pound-Drever-Hall frequency fluctuation sensing arrangement.	45
4.8	Sample Bode plots for frequency-tuning devices in a typical monolithic Nd:YAG laser.	46
5.1	Schematic representation of the FCS design process.	55
6.1	Multiple-stage sensor configuration.	64
6.2	Example of lower bound of open-loop gain for laser frequency noise suppression.	69
6.3	Example of "optimum" open loop gain.	70
6.4	Concept of induced lock acquisition arrangement.	77
6.5	Use of a network analyzer for measuring the frequency response of an amplifier stage.	80
6.6	Modified version of Fig. 2.1, relevant to troubleshooting and testing FCS performance.	82
6.7	Build-up of oscillation as a result of closed loop instability, seen at the sensor output.	84
6.8	Example of system that goes unstable when the overall gain is either too high or too low.	85
6.9	Sensor output for insufficient open-loop gain.	87
6.10	Lag-lead circuit used to increase open-loop gain at low frequencies.	88
6.11	Time record of actuator driver output $e_c(t)$ for a situation where tracking fails as a result of insufficient actuator range.	89
6.12	Use of an external sensor for testing tracking performance.	94
7.1	Diagram of Fig. 2.1, modified for the discussion of multiple control paths.	100
7.2	Modified version of the laser frequency noise suppression arrangement shown in Fig. 4.5.	101
7.3	Example of upper limit to the spectrum of frequency fluctuations in a free-running laser.	103
7.4	Closed-loop system with two parallel signal paths.	104

List of Figures xiii

7.5	Example of possible Bode diagram for the system of Fig. 7.4.	105
7.6	Block-diagram illustrating the concept of low-frequency gain boost.	106
7.7	Effect of low-frequency gain boost on overall loop gain.	107
7.8	Arrangement for suppressing mirror vibration caused by ground motion.	109
7.9	Example of displacement fluctuation spectrum affecting the position of the mirror of Fig. 7.8.	110
7.10	Nested loop arrangement for preventing actuator saturation.	111
7.11	Transformation of parallel path diagram into equivalent nested diagram.	112
7.12	Transformation of parallel gain-boost diagram into equivalent nested diagram.	112
7.13	PID compensator.	115
8.1	Modified version of Fig. 2.1, emphasizing the digital compensator.	120
9.1	Diagram of Fig. 2.1, shown with $E_r = 0$, for the discussion of active null measurements.	138
10.1	Diagram of control system with two sensors for the same variable.	142
10.2	Example of two-sensor arrangement for monitoring the motion of a system.	142
10.3	Equivalent sensor noise for two-sensor arrangement.	145
11.1	Example of actuator moving a mirror attached to a flexible structure.	148
11.2	Frequency response of PZT/mirror system in the presence of structure flexibility	149
11.3	Effect of structure flexibility on system stability	150
11.4	Decoupling of PZT/mirror system from structure resonances by momentum compensation.	151
A.1	RC network used to illustrate the concept of pole.	153
A.2	Bode plot for the RC network of Fig. A.1.	155
B.1	Inverting amplifier using an idealized operational amplifier.	160
B.2	A more realistic representation of an amplifier using an opamp.	161
B.3	Relationship between amplifier gain and bandwidth.	163
B.4	Examples of Bode diagrams for two opamps.	164

B.5	Model of feedback amplifier driving a capacitive load.	166
B.6	Effect of capacitive output loading on the open-loop gain of the feedback amplifier.	167
B.7	Basic photodiode circuit and photodiode model.	168
B.8	Transimpedance amplifier used as photodiode preamplifier.	168
B.9	Effect of capacitive input loading on the open-loop gain of the feedback amplifier.	169

Preface

> *You don't have to be a rocket scientist to be a rocket scientist.*
>
> David Letterman

The shelves of technical libraries hold hundreds, possibly thousands of titles covering this or another aspect of feedback control systems (FCS). Yet we found it hard to answer the question "Can you recommend a quick, practical introduction to control systems?" This kind of inquiry often comes from an audience consisting of project-minded people like engineers, scientists and graduate students, who have the following traits in common:

- A need to design or improve a complex product or instrument.

- A need to include a FCS in their product or experiment.

- A need for the control system to work well enough to ensure adequate functionality of the overall system. While this may sometimes require extreme levels of FCS performance, there is no particular desire to optimize performance in a rigorous mathematical sense. **Success is achieved when the product or the experiment perform as specified**.

- These people lack the time for pursuing the rigorous mathematical aspects of FCSs or a formal FCS education. Typically, in order to accommodate the needs of a project, they have to grow from novice to hands-on proficient in a few weeks or months.

In order to best serve the needs of our user constituency, this text sticks to a direct **hands-on** approach to FCS design. A handful of introductory level concepts related to differential equations and Laplace transforms are the only pre-requisites. To further facilitate the use of the book, a procedure/checklist style has been adopted wherever possible (including in this Preface!).

Keeping in mind the fact that the audience for this work sees the FCS as one piece of a larger puzzle where the aim is to achieve a specified level of performance for the overall system, we have chosen to expand the concept of feedback control system design by including:

- **The Big Picture**. If the need for a feedback control system is anticipated, this should be kept in mind when designing the other subsystems of the system under consideration. This will help avoid inconsiderate choices which may eventually result in extreme demands on the FCS. For example, if the need for actively damping a structure is expected, the structure should be designed with as much passive damping as possible, thus leading to relatively easy requirements on the FCS providing the active damping.

- Prototyping.

- Troubleshooting.

- Design iteration.

- A success criterion based on achieving a prescribed level of performance for the overall system.

In order to maximize accessibility and practical usefulness, we decided to keep the text short. As a consequence, we chose to forego presenting to the reader the rich variety of methods generated in the field of controls over the last 50 years; this book is based exclusively on frequency domain design methods. We also assumed that the reader has been exposed to Laplace transforms and related mathematical topics at a most basic level. For most effective use of this book and for improved productivity in designing feedback control systems that work, we recommend the following companions:

- The recently published comprehensive book by Lurie and Enright[1] for ultimate in-depth knowledge of frequency domain design of FCSs.

---

[1] B. J. Lurie and P. J. Enright, Classical Feedback Control, Marcel Dekker Inc., 2000.

- The excellent monograph by Oppenheim and Willsky[2] contains everything most readers will ever need to know about signals, sampling and various transformations connecting the time domain with the frequency domain, e. g. Laplace transforms.

- An introductory level text on control systems, like the one by DiStefano et al.[3]

Those interested in a historic collective appearance of many of the concepts mentioned in this book are encouraged to peruse the feedback amplifier design treatise by Bode.[4]

In the same spirit of keeping things short, the book does not expand on the subject of control systems simulation by computer. However, numerous frequency and time domain examples are used to illustrate the concepts and methods which are discussed. All simulation results shown here were generated using **Matlab**$^{(R)}$ by Matworks, Inc. Proficiency with this software package is highly desirable.

Everything in nature, and FCSs are no exception, is nonlinear in its response to stimuli. Frequency domain design methods, however, are based on the assumption of fully linear behavior. The book by Lurie and Enright is a good reference for design methods which take nonlinearity into account in a systematic way. While we feel that including nonlinear design methods is beyond the scope of this book, FCS design as described here incorporates an effective compromise consisting of:

1 **Linearity assumption**
It is assumed that when the feedback loop is closed, the system under consideration operates in a purely linear regime; frequency domain design methods are therefore applicable.

2 **Awareness of dominant nonlinearities**
Strong nonlinearities like

- Sensor range limits
- Actuator range limits
- Amplifier saturation close to the power supply rails
- Slew rate limitation

---

[2]. V. Oppenheim, A. S. Willsky, Signals and Systems, Prentice Hall, 1983
[3]J. J. DiStefano III, A. R. Stubberud, and I. J. Williams, Feedback and Control Systems, Schaum's Outline Series, McGraw-Hill, 1990.
[4]H. W. Bode, Network Analysis and Feedback Amplifier Design, Van Nostrand, NY, 1945.

are listed and used to determine the range in parameter space where the system can be considered linear.

3. **Lock acquisition mechanism**
   If the system "wanders" out of linear territory, the feedback loop may cease to function, a condition described as **loosing lock**. In order to restore loop operation, a **lock acquisition** arrangement needs to be added to the system, providing the following functions:

   - Detection of an out-of-lock condition.

   - Feeding an appropriate parameter bias to the actuator(s), in order to induce the system to re-enter the linear domain.

   - Holding actuator bias when feedback loop operation has been restored.

Many applications require that several sensors and actuators be used in the same system. We present the reader with a powerful and easy to use technique for designing systems with more than one actuator, in Chapter 7, or more than one sensor, in Chapter 10. While this technique is probably not adequate for dealing with complex acoustic noise reduction applications which require the simultaneous use of many microphones and loudspeakers, it does help to overcome the limitations of single sensors or actuators and enables the design of very high performance FCSs.

The unfortunate tendency of FCSs to go unstable is the reason why much of the work on control systems and many textbooks on the subject focus on the stability issue. Therefore, going to the library is of little help when a feedback control system problem arises and needs to be resolved in less time than it takes to go back to school and take a number of relevant classes. In order to provide some relief for the needy practitioner, the present guide will trade generality for simplicity and rigorous optimization for the pursuit of "good enough" performance. Mathematics will be minimized. The concept of "just-as-much-as-needed" is in fact extended to every aspect of the text and, by extension, to the design process itself. It turns out that this **intuitive** approach can result in FCSs with close-to-optimum performance, for many practical situations in the lab. In order to support trouble-shooting and refining of a FCS, once it has been built, this guide addresses in detail the practical aspects of building a working system.

Preface xix

In order to answer the FCS needs of this audience, the contents of this book have been divided into four distinct parts:

1. An elementary introduction to feedback control systems for the novice (Chapters 1-4).

2. A detailed description of a step-by-step FCS design process, used by the authors over many years. A handful of methods, easy to understand and use, are included here (Chapters 5-8).

3. Several applications of the concepts developed in the book are presented under the common heading **Special Topics** (Chapters 9-11).

4. Auxiliary material supporting the design process has been compiled in the **Appendices** (Appendix A-C).

Reading the material should be helped by a **Summary of Notations** and a **Summary of Examples**, both on page x.

Chapter-by-chapter, the material is organized as follows:

- Chapters 1, 2, which introduce some FCS essentials, and Chapter 3, which presents a discussion of closed-loop stability, are meant for the novice. Readers with some introductory knowledge on feedback control systems may cruise through Chapter 2 to familiarize themselves with the formulae, then proceed to Chapter 4.

- Chapter 4 illustrates the ingredients of FCS design via two examples:
    - An optical system for aircraft tracking. Concepts like sensor range and lock acquisition are explained here.
    - Frequency stabilization of a laser, where some important but often overlooked concepts, like in-band and out-of-band frequency ranges are introduced.

  These examples are frequently referred to later. The reader is encouraged at least to rush through this chapter.

- The following four chapters contain the main part of the book:
    - Chapter 5 contains a detailed discussion of the design approach advocated here. One important aspect is that prototyping, experimental testing and iteration are considered, by definition, integral to the design process. There are several advantages to this extended definition of design:

1 The broader systems level view removes constraints endemic to a compartmentalized approach, making the process easier and ultimately ensuring better overall performance at a lower cost.
2 It encourages direct experimental validation of "good enough" transfer functions and lock acquisition provisions.
3 Noise and other real-life implementation problems become apparent at an early stage, which leads to shorter concept-to-product cycles.

- Chapter 6 goes through the steps involved in the design of an FCS, including evaluating and trouble-shooting the prototype (Section 6) and refining the design in order to achieve the specified performance.
- Chapter 7 introduces multiple signal paths as a way to enhance system performance. Three useful configurations are discussed and compared starting from real life examples. A powerful practical stability criterion for this type of arrangements is introduced.
- Chapter 8 presents a recipe for specifying the components of a digital compensator and a short discussion of circumstances where it is advantageous to go digital.

- Part III addresses the following special topics:

  - Chapter 9 contains a discussion of how feedback control systems can be used to carry out active null measurements. These are measurements where the unknown value of a variable is determined by nulling it. The nulling effort is a measure of the unknown variable.
  - Chapter 10 is an extension of Chapter 7, which treats arrangements with multiple actuation paths, to the parallel use of two sensors for measuring the same variable.
  - Chapter 11 offers a discussion of stability problems caused by flexible elements in the plant and indicates how this systems can be stabilized.

- Some items related to the contents of this guide but outside its main logic flow are discussed in the Appendices.

  - Appendix A explains the effects of poles and zeros on the magnitude and phase of the transfer function. An electrical network representation of a pole is used for illustration.

Preface

- Appendix B discusses the stability of operational amplifiers used as photodiode preamplifiers and for driving capacitive loads.
- Appendix C addresses the limitations on signal knowledge which arise from the existence of quantization steps inherent in the analog-to-digital conversion of data samples. Ways of overcoming the least significant bit resolution limit are discussed.

In spite of our extensive efforts to weed errors out, some may have evaded detection. We are thus encouraging the readers to let us know about any kind of errors or suggestions for improving the book. Please contact us by e-mail at *Alex.Abramovici@jpl.nasa.gov*. This kind of help will be greatly appreciated.

Acknowledgments

We wish to thank Lawrence Chu (Intel Corp.), whose insistent "why" questions prompted us to engage in this project. We are grateful to Drs. Greg Neat and Lisa Sievers, both with the Jet Propulsion Laboratory, for their help on some of the technical aspects covered in the book. We are indebted to Drs. Serge Dubovitsky (Jet Propulsion Laboratory), Mark Milman (Jet Propulsion Laboratory), Carl Nash (formerly with the National Transportation Safety Administration) and Keith Riles (University of Michigan) for their insightful comments. Several stimulating discussions with Dr. Dan DeBra (Stanford University) have influenced the contents and the style of this book.

Our deep gratitude goes to Drs. Frank Dekens (Jet Propulsion Laboratory), Eric Gustafson (Stanford University), Peter Saulson (Syracuse University) and Benno Willke (University of Hannover), who virtually β-tested early versions of the manuscript and provided us with invaluable criticism and advice reflecting the user's point of view.

I

INTRODUCTION TO FEEDBACK CONTROL SYSTEMS

Chapter 1

CONTROL SYSTEM DIAGRAMS

Experimental investigation and engineering work often involve forcing a parameter, which is otherwise undetermined or does not have the desired value, to track a reference to a specified degree of accuracy. A common approach is to use a feedback control system (FCS). The ideas involved are illustrated in Fig. 1.1 on p. 5. Some physical process is taking place in the system of interest, called the **plant**. Under ideal conditions, in the absence of control the output of the plant would be constant. In reality, because of changes in temperature and other environmental parameters, and because of noise intrinsic to the plant, the output changes with time. Generally, the overall plant output can be regarded as a sum between the "pure" (and steady) output and a **disturbance**. Because of the fluctuating disturbance, the plant output is designated as **free-running**. In practice, the "pure" plant output cannot be measured independently of the disturbance. In fact, the only measurable quantity at the plant output is the free-running output, and therefore the distinction between the "pure" output, the disturbance and the free-running output is purely academic.

As illustrated in Fig. 1.1b, output control is effected as follows:

1 The output is measured using a device designated as **sensor**. The output of the sensor is typically a voltage which depends on the plant output in a known way. It is desirable to have a sensor with linear, or close to linear characteristic.

2 The sensor output is compared to a reference level by generating their difference at the leftmost summing node in Fig. 1.1b.

3 The difference between the reference level and the sensor output, is called the **error signal**. This is an electrical signal, which is filtered and amplified by the **compensator**, also called **controller**.

4 The compensator output is used to drive the correction device, called **actuator**. The actuator converts the electrical signal at its input into a signal which can directly modify the plant, and thus its output.

5 When the sensor output is close to the reference signal, the error signal is small and thus the correction applied by the actuator to the plant becomes small as well. This happens when the controlled plant output has approximately the required value. The closed-loop system thus is capable of automatically adjusting the correction to the amplitude of the difference between the desired output and the actual output.

The arrangement of Fig. 1.1b has several noteworthy features:

- Output control occurs by feeding the output back into the plant via the sensor, the compensator and the actuator, hence the names **feedback control system** or **closed-loop control system** for this kind of setup.

- The correction signal is obtained by processing the error signal, which is the difference between the reference input and the sensor output. Thus, changing the reference signal provides the operator with a way of modifying the output. This is why a **time-dependent** reference signal is sometimes called **control input** or **command**.

- The purpose of building the control system is disturbance suppression and output control. Part of the price one pays for these benefits is added complexity. Moreover, the added components provide additional access paths over which other disturbances, referred to as **noise**, can be added to the output.

- Finally, the closed loop configuration is prone to instability, and it takes a certain amount of design effort to ensure stable operation.

All the above points will be addressed in detail in this book.

The diagrams of Fig. 1.1 can be viewed as representing the flow of signals along the arrows. Each block then represents a signal processing function. For linear systems as considered in this book, the processing function carried out by each block is described by a differential equation,

Control System Diagrams

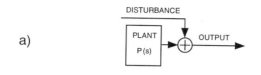

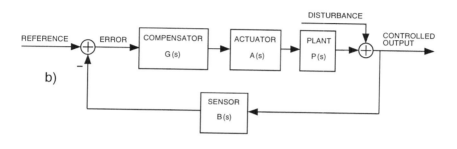

Figure 1.1. a) Schematic representation of a plant that is not controlled, b) diagram representative of many commonly encountered feedback systems used to control plant output. The "-" sign below the left-most summing junction indicates that the sensor output is first sign-reversed, then added to the input. $B(s)$, $G(s)$, $A(s)$, $P(s)$ represent the transfer functions of the sensor, compensator, actuator and plant, respectively.

while the input signal to a block acts as a forcing function. An alternative time-domain description of a system is its **impulse response**, which is the solution of the corresponding differential equation when the forcing function is the delta-function $\delta(t)$. The output of a system is given by the convolution product between the input signal and the impulse response of the system. For anything but the simplest systems, calculating the output signal by solving the differential equations or by using the convolution of the input with the impulse response is not practical for the FCS design process. Significant simplification in deriving the output signal is brought about by the use of the Laplace transform, which associates a function $a(t)$ with its Laplace transform $A(s)$, where

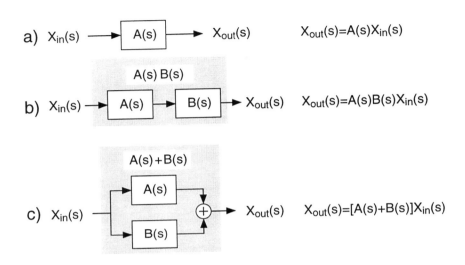

Figure 1.2. Algebraic meaning of blocks in feedback control system diagrams.

$s = \sigma + j\omega$ and $\omega = 2\pi f$.[1] The Laplace transform of the output of a system is obtained simply by multiplying the Laplace transform of the input signal, i. e. the Laplace transform of the forcing function, with the Laplace transform of the system impulse response, called **transfer function**. If $A(s)$ is the transfer function of a system, $A(j\omega)$, obtained by setting $\sigma = 0$, is called the **frequency response** of the system. $A(j\omega)$ is the complex gain at the frequency $f = \omega/2\pi$ and is thus a convenient frequency domain description of the system.

Diagrams like the one in Fig. 1.1 are particularly useful since they can be endowed with an algebraic meaning, as shown in Fig. 1.2:

- Each block represents a transfer function.

---

[1] The definition of the Laplace variable, the Laplace transform and of other transformations between the time domain and the frequency domain can be found in many textbooks, e. g. A. V. Oppenheim, A. S. Willsky, Signals and Systems, Prentice Hall, 1983

Control System Diagrams

- The signal flow in and out of the blocks is indicated with arrows. Just the arrow itself represents a unity transfer function.

- The output signal of a particular block is the input signal for the block that follows.

- As shown in Fig. 1.2a, the output signal is obtained by multiplying the input signal with the transfer function associated with the block.

- The transfer function corresponding to two blocks in series is equal to the product of the transfer functions associated with the two blocks (Fig. 1.2b).

- The transfer function corresponding to two blocks in parallel is equal to the sum of the transfer functions associated with the two blocks (Fig. 1.2c).

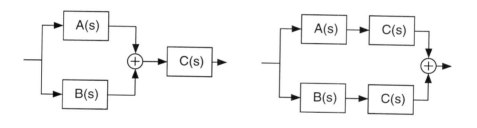

Figure 1.3. Example of equivalent blocks.

Fig. 1.3 shows two diagrams which according to the above rules represent the same transfer function $[A(s) + B(s)]C(s)$. The usefulness of this kind of diagram manipulation will become apparent later, in particular in Chapter 7.

Chapter 2

SIGNALS IN THE FEEDBACK LOOP

As already mentioned, this book promotes the idea of using just the right amount of mathematics in the design process. A few indispensable formulae will be derived in this Chapter; the rest of the necessary math will be delivered in small portions as the need arises.

1. RELATIONSHIPS BETWEEN SIGNALS IN A LOOP

The discussion is based on based Fig. 1.1, reproduced below for convenience.
Fig. 2.1 is a modified version of Fig. 1.1, showing the signals and the noise contributions. To bring the discussion one step closer to reality, Fig. 2.1 shows that the actuator receives its signal from a **driver**, which in many cases is just a power amplifier.

In order to reduce all calculations to simple algebra, the considerations in this Chapter are using the Laplace transforms of various quantities of interest. With this in mind, all variables are denoted by upper-case symbols and the s-dependence is no longer explicitly shown.

At first sight, following the signals through the diagram of Fig. 2.1 may appear confusing, due to the closed-loop character of the system. However, understanding is greatly helped by the fact that working in the frequency domain or in the s-plane is equivalent with assuming a steady-state condition. Following the arrows in Fig. 2.1, one notes that the

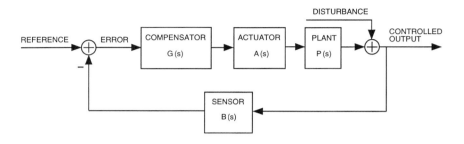

Fig. 1.1

output signal X_{Cl} is converted to an electrical signal $E_s = B \cdot (X_{Cl} + N_s)$ by the sensor. It is worth noting that the sensor contributes a noise term N_s. The difference $E_r - E_s$, called **error signal**, formed at the leftmost summing junction in Fig. 2.1, is a measure of how far off the output is from the desired value. In addition to sensor noise included in E_s, two error terms are affecting this signal:

- N_r, reflecting the imperfect nature of the standard from which the reference signal is derived.
- N_G, describing electronics noise in the compensator.

The compensator is essentially a filter with transfer function $G(s)$. The Laplace transform of the voltage at the compensator output is $E_G = G(E_r - E_s + N_r + N_G)$. This signal is further amplified and filtered by the actuator driver, then converted by the actuator into the correction signal $X_c = AHE_G$. The correction signal is multiplied by the plant transfer function P and is added to the free-running variable,[1] yielding the closed-loop output $X_{Cl} = X_{fr} + X_c$. Eliminating the intermediate variables E_s and E_G from the above equations yields:

$$X_{Cl} = \frac{(E_r + N_r + N_G)}{B} \cdot \frac{L}{(1+L)} - N_s \frac{L}{(1+L)} + X_{fr} \frac{1}{(1+L)} \quad (2.1)$$

---

[1] In the absence of correction, $X_{fr} = X + X_d$.

Signals in the Feedback Loop

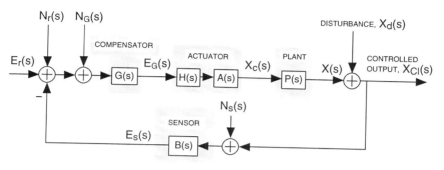

Figure 2.1. Schematic representation of a feedback control system, showing the signals and the noise contributions. The FCS components are represented by their transfer functions, and the signals by their Laplace transforms. $X_{Cl}(s)$: output of controlled plant; $X_c(s)$: correction signal; $E_G(s)$: compensator output; $N_G(s)$: error in the electronics, referred to compensator input; $E_r(s)$: reference voltage, or command signal; $N_r(s)$: reference error; $E_s(s)$: sensor output; $N_s(s)$: sensor error. $P(s)$ is the plant transfer function. The actuator consists of a driver with transfer function $H(s)$ and the actuator itself with transfer function $A(s)$. $G(s)$ is the compensator transfer function, and $B(s)$ is the transfer function of the sensor. G and H are measured in Volts/Volt, A is measured in units(plant input)/Volt, and B is measured in Volts/units(X). P is measured in units(X)/units(plant input). It is customary to distinguish between different contributions to a certain type of error by their frequency range. A **DC** error is called **offset**, a **slow** undesirable change occurring over timespans of minutes or longer is called **drift**, and faster spurious changes are referred to as **noise**.

where $L = BGHAP$ is called **open-loop transfer function** . This is because it is obtained by breaking the loop in Fig. 2.1 at some arbitrary point, then following the arrows and multiplying the Laplace transforms of the various elements as they are encountered. Disregarding sensor noise, the ratio between the sensor output BX_{Cl} and the control input E_r, with the loop closed, is called **closed-loop transfer function** for the control input. It can be obtained from Eq. 2.1 by setting $N_s = 0$, $N_r = 0$, $N_G = 0$, and $X_{fr} = 0$:

$$\frac{BX_{Cl}}{E_r} = \frac{L}{1+L} \qquad (2.2)$$

For $|L| \gg 1$, Eq. 2.1 reduces to:

$$X_{Cl} \simeq \frac{1}{B}(E_r + N_r + N_G) - N_s + \frac{X_{fr}}{L} \qquad (2.3)$$

while Eq. 2.2 becomes $BX_{Cl}/E_r = 1$. Equations 2.2 and 2.3 can be interpreted as follows:

- With the loop closed, the main contribution to X_{Cl} is E_r/B, proportional to E_r. Division by the sensor gain B merely scales the control input and converts it into units of X, i. e. it leads to a comparison of "apples to apples" between the plant output X and the control input E_r.

- With the loop closed, the contribution of the the free-running plant output (which includes the disturbance) to X_{Cl} is X_{fr}/L. The free-running output is thus suppressed by the magnitude of the open loop transfer function $|L|$. In other words, the higher the required accuracy for tracking the control input E_r, the higher the necessary open-loop transfer function magnitude $|L|$.

- Closing the loop as shown in Fig. 2.1 adds to the output X_{Cl} the noise associated with the sensor, the reference input and the compensator. Noise from the actuator has been disregarded because it is assumed that compensator noise dominates over actuator noise. This should always be the case since actuator output noise competes with reference noise and compensator noise, both multiplied by the transfer function GHA of the previous stages of the loop.

In conclusion, closing the FCS loop causes X_{Cl} to track the reference scaled by the sensor gain, E_r/B, to accuracy ($X_{fr}/|L|$ plus noise in the system) and suppresses the free-running plant output to X_{fr}/L, which illustrates the usefulness of FCSs in tracking applications, in particular when high $|L(s)|$ or **high open-loop gain** $|L(j\omega)|$ and low noise can be obtained. This is also why **performance** and high open-loop gain are synonymous in the FCS context.

In the high gain limit ($|L(j\omega)| \to \infty$), the correction signal is calculated from from Eq. 2.1:

$$PX_c = X_{Cl} - X_{fr} = \left[\frac{1}{B}E_r - X_{fr}\left(1 - \frac{1}{L}\right)\right] + \left[\frac{1}{B}(N_r + N_G) - N_s\right] \qquad (2.4)$$

Signals in the Feedback Loop 13

where only terms no smaller than $\propto 1/L$ have been kept for each contribution. Again, we find that the correction removes X_{fr} to accuracy $1/L$, and replaces it with the control input scaled by the sensor gain, E_r/B, and noise. In the high-gain limit, the sensor output is:

$$E_s = B\left(X_{Cl} + N_s\right) = \left[E_r + \frac{B}{L}X_{fr}\right] + \left[N_r + N_G + \frac{B}{L}N_s\right] \quad (2.5)$$

It follows that compared to the open-loop case, where $E_s = B(X_{fr}+N_s)$, the X_{fr} contribution to the closed-loop sensor output is attenuated by a factor equal to the open-loop gain $|L(j\omega)|$. A remarkable feature of Eq. 2.5 is that the contribution of the sensor noise to the sensor output is also attenuated by the open-loop gain, compared to the open-loop case. Thus, sensor noise will affect the controlled output as shown by Eq. 2.3, but it will appear attenuated at the sensor output. Therefore, when it comes to evaluating tracking performance, it is a good idea to consider an additional sensor which is not part of the FCS loop. This issue will be discussed in detail in Section 6.4 of Chapter 6.

The main result of this Subsection can be stated as follows:

For high open-loop gain $|L(j\omega)|$ and low noise, the FCS converts the free-running variable, including the disturbance, into the reference value for the variable X, in other words it ensures proper tracking of the reference. See Eqs. 2.3, 2.4 and 2.5.

---

2. SIMPLIFIED FCS DIAGRAMS

For some of the considerations in this book, Fig. 2.1 is too detailed, and the issues being discussed can be better emphasized by using simplified versions. Simplifications include:

- Disregard all noise contributions.

- Assume that the high gain limit $|L(j\omega)| \gg 1$ applies.

14   FEEDBACK CONTROL SYSTEMS

- Disregard the different physical nature of the various signals, so that all signals are expressed in units of X. For example, one would set $B = 1$, so that $E_s = X_{Cl}$ and $X_r \stackrel{\text{def}}{=} E_r$.
- Lump all the elements of the FCS into a single block carrying the entire open-loop transfer function $L(s)$ or open-loop frequency response $L(j\omega)$.

The simplified forms of Eqs. 2.3, 2.4 and 2.5 are:

$$X_{Cl} = \frac{E_r}{B} + \frac{X_{fr}}{L} = X_r + \frac{X_{fr}}{L} = X_r \qquad (2.3a)$$

$$PX_c = \frac{E_r}{B} - X_{fr} = X_r - X_{fr} \qquad (2.4a)$$

$$E_s = X_r + \frac{X_{fr}}{L} \qquad (2.5a)$$

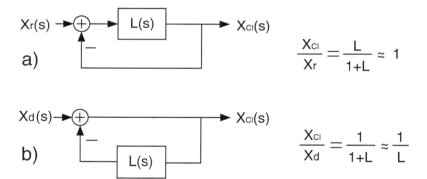

Figure 2.2. Simplified feedback control system diagrams at the limit $|L(j\omega)| \gg 1$. a) Reference-following (tracking) system. b) Disturbance suppression system.

The diagrams of Fig. 2.2 are examples of simpler versions of Fig. 2.1, which are useful in explaining two widespread uses of FCSs. Fig. 2.2a represents a system which forces the output to track the reference/control input X_r. This diagram includes the additional simplification that the disturbance X_d is not shown. This is justified by the fact that, according to Eq. 2.3, the free-running output X_{fr}, which includes the disturbance, is suppressed by the loop gain. The relevant closed-loop transfer function is $L/(1+L) \approx 1$. Another important case is where one is interested in disturbance suppression, but not in tracking a reference. Then one can set $X_r=0$ and distort the diagram until it looks like Fig. 2.2b. The relevant closed-loop transfer function, the ratio X_{Cl}/X_d, is $1/(1+L) \approx 1/L$.

3. ACTUATOR OUTPUT RANGE

Disregarding noise and assuming high gain so that $L/(1+L) \approx 1$, Eq. 2.4 on p. 12 reads $X_c = (E_r/B - X_{fr})/P$. This means that the range of the actuator has to be wider or at least equal to the range of departure of the free-running variable from the reference, divided by the plant transfer function. If this condition is not met, the actuator goes into a state called **saturation**. In this state, the system can no longer apply the appropriate correction signal and closed-loop operation breaks down. Actuator saturation is one case of nonlinear behavior which has detrimental effects on adequate closed-loop operation. Rigorous treatment of nonlinearities and their effects is beyond the scope of this book. The approach recommended here is to avoid the onset of actuator saturation by choosing a device with sufficient range. When it is just not possible to choose a single actuator with sufficient range, one can add a control path with a second actuator. This solution for the actuator saturation problem is discussed in detail in Chapter 7.

Chapter 3

STABILITY

The price one has to pay for the benefits of closed-loop operation includes complexity and the potential for instability. The latter manifests itself by causing oscillation and possibly amplifier or actuator saturation. Individually or together, these effects prevent the system from functioning as intended. This Chapter offers a brief overview of stability and instability. Section 1 introduces the reader to the concept of stability in the time domain using differential equations. Section 2 presents the Nyquist stability criterion, which in this book is the guiding principle for the part of FCS design aiming at ensuring stable operation. Section 3 offers some intuitive insight into the build-up of instability.

1. DIFFERENTIAL EQUATIONS, LAPLACE TRANSFORMS AND STABILITY

The dynamics of many systems of interest can be described by differential equations; this allows stability to be defined and described in a precise mathematical language. In general, the evolution of a parameter $x(t)$ associated with a linear system is described by the nth order differential equation with constant coefficients:

$$a_0 \frac{d^n x(t)}{dt^n} + \frac{d^{n-1} x(t)}{dt^{n-1}} + ... + a_n x(t) = f(t) \qquad (3.1)$$

where $f(t)$ is the forcing function. The general behavior of $x(t)$ is governed by the solution of the homogeneous counterpart of Eq. 3.1, i.e. by the solution corresponding to $f(t) = 0$, which can be written as:

$$x_h(t) = \sum_{i=1}^{n} c_i e^{r_i t} \qquad (3.2)$$

where r_i are the roots of the characteristic equation associated with Eq 3.1:

$$\sum_{k=0}^{n} a_k r^k = 0 \qquad (3.3)$$

The sum on the left-hand side of Eq. 3.3 is called $P_n(r)$, the characteristic polynomial associated with Eq. 3.1. When multiple roots occur, Eq. 3.2 has a slightly different form, but is still a sum of exponentials. A system is stable when $x(t)$ or any other function which describes the system is bounded as $t \to \infty$. It is worth noting that for complex roots $r_i = \sigma_i + j\omega_i$, the real part represents a real exponential, while the imaginary part describes an oscillatory behavior. In other words: **a system is stable when the real parts of the roots of the characteristic polynomial are all negative or zero**. This ensures that the real exponentials in Eq. 3.2 do not become unbounded. An equivalent statement is that the characteristic polynomial should have no roots in the right-hand side of the s-plane, where $s = \sigma + j\omega$.

In order to translate the above formulation of the stability criterion into common control system language, note that the Laplace transform $X(s)$ of the solution of Eq. 3.1 is:

$$X(s) = \frac{F(s)}{P_n(s)} \qquad (3.4)$$

Since the output $X(s)$ is obtained by multiplying the input $F(s)$ with the expression $1/P_n(s)$, the latter is the transfer function of the system described by Eq. 3.1. In terms of $1/P_n(s)$, the condition that there be no roots of Eq. 3.3 in the right-hand side of the s-plane translates into the requirement that the transfer function have no poles in the right-hand side of the s-plane. As shown in Fig. 2.2 on p. 14, when a feedback control system is built by closing the loop, the transfer function of the new system is $L(s)/(1+L(s))$, or $1/(1+L(s))$, depending on what task the system is designed to perform, $L(s)$ being the open-loop transfer

Stability

function of the system. In either case, the denominator is the same, $(1 + L(s))$, and the stability condition formulated above reads:

> For the closed-loop system to be stable, $[1 + L(s)]$ should have no zeros in the right-hand side of the s-plane.

An electrical network realization of poles and a discussion of frequency response properties related to poles and zeros are given in Appendix A.

2. THE NYQUIST STABILITY CRITERION AND BODE DIAGRAMS

Much of the classic feedback control system research revolved around finding conditions which L has to satisfy in order to ensure that $1 + L(s)$ has no zeros in the right-hand side of the s-plane. Since these conditions would guarantee stable behavior for the closed-loop system, they are referred to as stability criteria. The **Nyquist criterion**[1] is among the most frequently used ones. A simplified form of the Nyquist criterion is given below.

With reference to Fig. 3.1 on the previous page, the Nyquist criterion states that a closed loop system with open-loop gain $L(j\omega)$ is stable if it is open-loop stable and if the plot of $\mathrm{Im} L(j\omega)$ versus $\mathrm{Re} L(j\omega)$ has no clock-wise encirclements of the (-1,0) point in the Nyquist diagram. Even if the Nyquist plot does not encircle the (-1,0) point clockwise and the system is thus theoretically stable, overshoot and ringing result upon applying a step function command at the input of the system, if the plot comes too close to the (-1,0) point. This leads to the concepts of **phase** and **gain margins**, illustrated in Fig. 3.2. Essentially, in order to keep overshoot and ringing below an acceptable level, i. e. in order to ensure **robust** system behavior, one needs to keep the Nyquist plot outside an

---

[1] see e. g. J. J. DiStefano *et all*, Feedback and Control Systems, Schaum's Outline Series, McGraw-Hill, 1990.

FEEDBACK CONTROL SYSTEMS

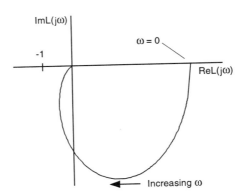

Figure 3.1. Example of a Nyquist diagram. According to the Nyquist criterion, the plot in this diagram corresponds to a stable system.

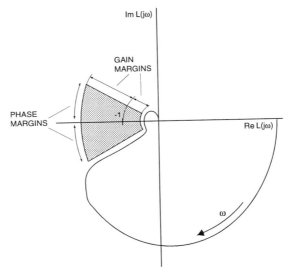

Figure 3.2. The concepts of phase and gain margins. As long as the locus of $\{\mathrm{Im}\,L, \mathrm{Re}\,L\}$ stays outside the shaded area, ringing and overshoot will not exceed a certain level.

Stability

area including the (-1,0) point. The wider that area, the more robust the system is.

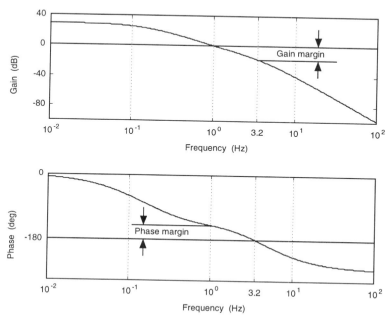

Figure 3.3. Example of Bode plot corresponding to a stable system. In this example, the unity gain (0 dB) frequency is 1 Hz and the phase margin is 38°. The phase becomes -180° at 3.2 Hz, and the gain margin is 17 dB.

A more convenient version of the Nyquist stability criterion can be formulated in terms of **Bode plots**. An example of Bode plot is given in Fig. 3.3, which represents the frequency response $L(j\omega)$ using two diagrams:

- A **log-log** plot of $|L(j\omega)|$ versus ω or f. In keeping with the traditions of electrical engineering, the quantity actually plotted is the gain in **dB**, i. e. $20\log_{10}|L(j\omega)|$. The main advantage of this representation lies in the fact that when the slope of the log-log magnitude plot is constant, the phase lag $\phi(L)$ is proportional to the slope:

$$\frac{\phi(L(j\omega))}{90^\circ} = \frac{\text{Slope}\,(|L(j\omega)|)}{20 \text{ dB/decade}} = \frac{\text{Slope}\,(|L(j\omega)|)}{6 \text{ dB/octave}} \qquad (3.5)$$

The Bode plots thus provide the designer with the convenience of having to keep in mind only two variables, frequency and gain, as opposed to gain, frequency **and** phase. This makes the design process simpler and more transparent.

Eq. 3.5 is a special case of **dispersion relations** which provide a Hilbert transform relationship between the real and complex part of certain complex functions describing physical properties of systems, in this case the transfer function.[2] It is worth noting that these relations are a mathematical expression of the fact that cause has to precede effect.

- Even though the magnitude plot contains the phase information, a plot showing the phase of the open-loop frequency response $\phi(L(j\omega))$ as a function of ω or f is often added for convenience. The phase is represented on a linear scale in degrees, while the frequency scale is logarithmic.

In terms of Bode plots, the Nyquist criterion requires that the system be open-loop stable and that $\phi(L) > -180°$ at frequencies where $|L(j\omega)| = 1$. In the language of Bode plots, the idea that system robustness requires some minimal distance between the $\{\text{Im}L, \text{Re}L\}$ locus and the (-1,0) point in the Nyquist diagram is again expressed in terms of gain margin and phase margin. For Bode plots which are smooth and monotonic around the unity gain frequency,[3] the stability margins can be defined as follows (see Fig 3.3):

- The gain margin in dB is $g_m = -20\log_{10}|L(j\omega_1)|$, where $\omega_1 = 2\pi f_1$ and f_1 is the frequency where $\phi = -180°$.

- The phase margin in degrees is $\phi_m = 180 + \phi(f_0)$, where f_0 is the frequency corresponding to $|L(j\omega)| = 1$.

Higher margins lead to a more robust system. However, according to Eq. 3.5 a higher phase margin means a shallower slope of $L(j\omega)$ around the unity gain frequency, which in turn means that the rate at which the gain can be ramped up towards lower frequencies is decreased. This sets a limit to the maximum achievable gain. Therefore, the price for excessive robustness is decreased performance.

---

[2] The reader interested in learning more about Hilbert transforms and dispersion relations may peruse, for example, the book by H. M. Nussenzveig, Causality and Dispersion Relations, Academic Press, New York, 1972.
[3] which is true in most cases of interest

Stability

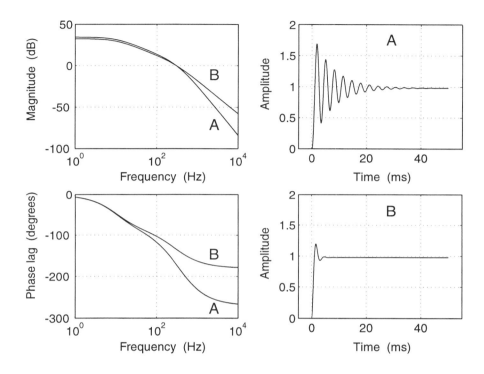

Figure 3.4. Connection between phase margin and time domain behavior of a closed-loop system. The Bode plots correspond to two open-loop frequency response functions with the same unity gain frequency f_o =300 Hz, but with different phase margins: ϕ_m=10° (A) and ϕ_m=45° (B). The traces on the right show the output $x_{Cl}(t)$ for a unit step function command. While Trace A shows substantial overshoot and ringing, Trace B is almost a faithful representation of the input command, a unit step function.

The connection between phase margin and the amount of overshoot and ringing in a system is illustrated in Fig. 3.4, for a reference tracking arrangement driven with a command signal equal to a unit step function. When the phase margin is $\phi_m = 10°$, dramatic ringing and overshoot are present (Trace A); the system is on the verge of oscillation. The main component of the ringing corresponds to the unity gain frequency of 300 Hz. When $\phi_m = 45°$, the output of the system tracks the input with little overshoot and almost no ringing (Trace B). A phase margin of 90° (not shown) corresponds to critically damped behavior. Thus, the higher the phase margin, the more resistant to oscillation the system is, as stated earlier on p. 19.

The above example shows that while 10° is not an adequate phase margin for a system tracking a command signal, 45°-60° is sufficient. In practice one finds that, for disturbance suppression applications, a lower phase margin $\phi_m = 30°$ is adequate. Given the conflict between the requirements for high open-loop gain on one side and high phase margin on the other side, the question arises of how to find the lowest phase margin which still ensures that ringing does not exceed a reasonable level. In principle, the lowest acceptable value of the phase margin has to be determined according to the specific requirements of each project. Alternatively, one can ask what is the highest slope of the frequency response which is still consistent with robust system operation. A practical way to address this question, based Bode's frequency domain method, is presented in the Chapter on system implementation.

The practical overshoot/ringing reduction approach taken here relies on the simple rules listed in the Box below.

Ringing Reduction Approach

- Require that the phase margin be at least 30°.
- If testing the system in the presence of signals typical for the loop under consideration shows that there is too much ringing and/or overshoot, redesign the loop to increase the phase margin.

We conclude this Section by introducing the **asymptotic Bode diagrams** illustrated in Fig. 3.5.

Asymptotic Bode diagrams are obtained as follows:

Stability

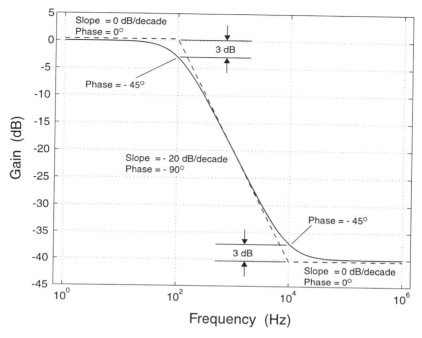

Figure 3.5. Example of asymptotic Bode diagram. The frequencies where two straight lines meet, 100 Hz and 10 kHz in this example, are called corner frequencies.

- Identify frequency spans where $|L(j\omega)|$ has constant slope. For the common case of rational transfer functions, the constant slope sections correspond to multiples of ± 20 dB/decade (the same as ± 6 dB/octave).

- Draw straight lines with slopes as above and with gains matching the corresponding constant slope portions of $|L(j\omega)|$.

- Extend the straight lines to obtain a continuous trace.

Asymptotic Bode diagrams are useful because they are easy to draw without using any computational help and because they are accurate for most frequencies. They depart form the correct values around frequencies where a transition between two values of the slope takes place, called **corner frequencies**, as shown in Fig. 3.5.

3. SYSTEM STABILITY: INTUITIVE APPROACH

The following discussion of stability will be based on the diagram of a disturbance suppression system, shown in Fig. 3.6. The sensor, the compensator and the actuator have all been lumped into one block described by L, the open-loop transfer function of the system. The corresponding simplified forms of Eqs. 2.1 and 2.4 are:

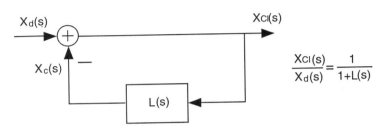

Figure 3.6. Diagram of feedback system for disturbance suppression, used as an example for stability discussion.

$$X_{Cl} = \frac{X_d}{1+L} \qquad (3.6)$$

$$X_c = \frac{X_d L}{1+L} \qquad (3.7)$$

The system of Fig. 3.6 will be tested for stability using the forcing function (disturbance) shown in Fig. 3.7.
This is the sinc function $(1/\pi t)\sin 2\pi BW t$, which has a flat spectrum with sharp cut-off:

$$X_d(f) = \begin{Bmatrix} 1 & |f| < BW \\ 0 & |f| > BW \end{Bmatrix} \qquad (3.8)$$

It is worth noting that the single sided frequency bandwidth BW is equal to the inverse of the pulse duration T, measured between the two zero points nearest to the central peak. The sharp cut-off in the spectrum of the test forcing function is convenient for exploring the connection

Stability

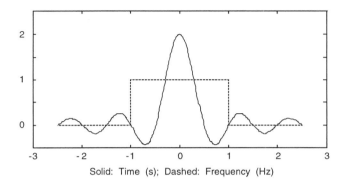

Figure 3.7. Sinc function disturbance. Solid: time dependence, dashed: spectrum. X-axis units for solid trace: time (s). X-axis units for dashed trace: frequency (Hz). The pulse shown has 1 s width, and 1 Hz bandwidth.

between stability and the properties of the system frequency response $L(j\omega)$ in arbitrarily chosen frequency bands.

The usual definition of stability will be used: a system is stable when it settles following a disturbance.

A particular open-loop frequency response $L(j\omega)$, with three poles at 0.1 Hz, a zero at 0.31 Hz and unity gain at 1 Hz, will be used as an example. This function, plotted in Fig. 3.8, has a phase lag of 180.1°, i. e. $\phi = -180.1°$ at the unity gain frequency of 1 Hz (or 2π radians/s). Thus, the closed loop system is expected to be unstable, according to the Nyquist criterion.

In the absence of feedback, this transfer function corresponds to a stable system. The onset of instability must then be connected with the presence of the correction signal, which is subtracted from the input signal in order to generate a null output. Inspection of Eq. 3.7 on p. 26 suggests that there are three distinct frequency regimes for the correction signal:

1. Below the unity gain frequency

2. At and just around the unity gain frequency

3. Above the unity gain frequency

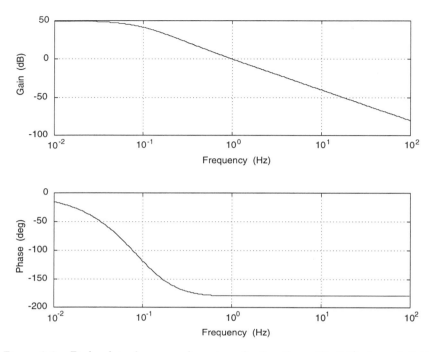

Figure 3.8. Bode plot of an open loop transfer function with 0.1° phase deficit, used to illustrate this discussion.

Frequencies Below the Unity Gain Frequency of $L(j\omega)$.
Well below the unity gain frequency, $|L(j\omega)|$ is high for the example discussed here (see Fig. 3.8). Thus, according to Eq. 3.7, the correction signal is approximately equal to the input signal; the input disturbance is successfully suppressed and there is no ringing at the output. In this case, the width of the test pulse would be larger than the inverse of the unity gain frequency, as its spectrum needs to drop to zero before unity gain is reached. Fig. 3.9 illustrates this point with a system response calculated using Matlab. Indeed, when a 4 s test pulse is fed into the system, pulse suppression and rapid settling is seen, even though, according to the Nyquist criterion, the system is expected to be unstable. According to Eq. 3.7, x_c is well behaved in this regime regardless of the phase of $L(j\omega)$. As long as the gain is high and the signal spectrum does not extend to the unity gain point, stable behavior is expected. This

Stability

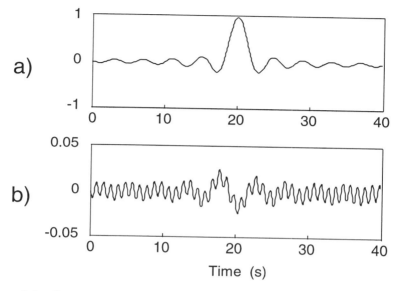

Figure 3.9. System response to a 4 s pulse (0.25 Hz bandwidth). a) input pulse; b) system output. Note that the pulse is suppressed 40 times, consistent with the open loop gain around 0.25 Hz (Fig. 3.8). Some 1 Hz ringing, excited by wideband noise due to the computation process, is present.

is of advantage when high gain is necessary at low frequencies, in spite of a relatively low unity gain frequency, which requires that the gain be dropped rapidly. Since each simple low-pass filter contributes 90° (see Appendix A), the phase lag can stack up quite high.

Frequencies Above the Unity Gain Frequency.
In this regime, $|L(j\omega)|$ is low, and the correction signal is small as well. Since little correction is applied to the input, the latter propagates through the system with almost no attenuation, as if there were no feedback. In this regime there is no need to worry about instability.

Frequencies at and Around the Unity Gain Frequency.
From Eq. 3.7 one finds that system behavior is likely to depend strongly

on the phase of $L(j\omega)$, because of the expression in the denominator, where a complex number with modulus close to unity is added to 1.

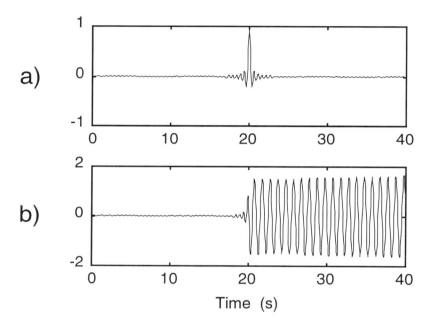

Figure 3.10. System response to a 0.5s pulse (2 Hz bandwidth). a) input pulse, b) system output. Note that the ringing is slowly building up, as expected with a very low phase deficit of 0.1° at the unity gain frequency.

- If the phase ϕ of $L(j\omega)$ is small, the correction signal will be smaller than and slightly out of phase with the input test pulse, but input attenuation will still occur, albeit less effectively. The test pulse will pass through with some attenuation, and the system will slowly settle.
- If ϕ is close to -180°, bad behavior is expected, as the denominator in Eq 3.7 tends to vanish. The key effect here is that the correction pulse becomes very large. With real systems, this is made worse by saturation in the amplifiers, which sometimes increases the phase shift. This, in turn, causes the correction pulse to miss the input

pulse at the correction point, so that two pulses are now traveling through the system. A third one is then generated, and there will be a never ending succession of pulses at the output, caused by just one input pulse. In other words, the system is unstable. The "test" pulse which adequately illustrates this regime has a spectrum extending beyond the unity gain frequency and duration less than the inverse of the unity gain frequency. Fig. 3.10 illustrates that the system does, indeed, display sustained high-amplitude ringing in response to such a pulse.

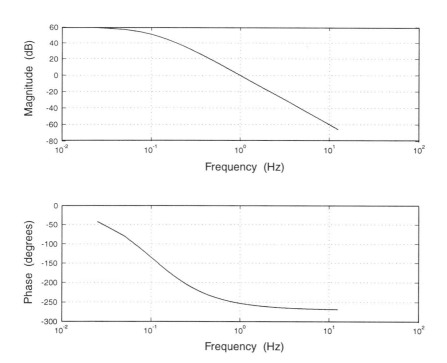

Figure 3.11. Bode diagram of system used to illustrate unstable behavior when the phase lag is substantially larger than 180°. In this example, $L(j\omega)$ has 3 poles at 0.1 Hz, unity gain frequency of 1 Hz, and approximately 253° phase lag at unity gain, i.e. a 73° phase deficit.

32   FEEDBACK CONTROL SYSTEMS

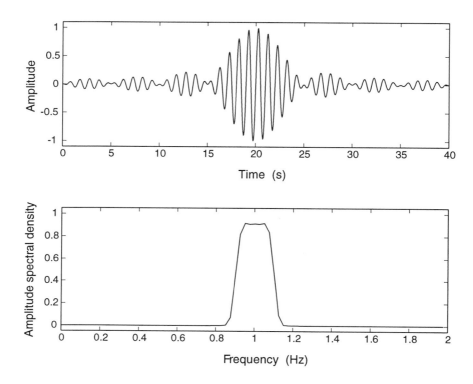

Figure 3.12. Input signal used to illustrate the effect of a large phase deficit at the unity gain frequency. a) input signal, b) spectrum of input signal, showing that most of the power is concentrated in a 0.2 Hz band around 1 Hz.

- If ϕ is well below -180°, for example $\sim -270°$, apparently no anomalous correction signal is generated, as the two terms in the denominator of Eq. 3.7 do not cancel. However, instability still takes place. In order to gain some insight why, consider the effect of the loop with 253° phase lag at the unity gain frequency (Bode plot in Fig. 3.11) on the input signal shown in Fig. 3.12. This signal was chosen because its spectrum is nonzero only in a narrow band around the unity gain frequency. After passing through the feedback network, the signal will be lagging 253° behind the input pulse. As illustrated in Fig. 3.13, this phase lag causes a delay which results in the correction signal missing the feature in the input signal which it is intended to correct. In fact, the heavily delayed correction ends up enhancing the input signal instead of suppressing it. This causes the output to build up to higher and higher amplitudes, in other words causes instability. Therefore, it is not necessary to have the denominator of Eq 3.7 vanish for instability to occur; any phase lag larger than 180° would do.

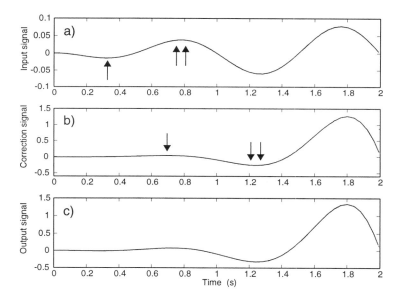

Figure 3.13. Instability build-up for the system with frequency response shown in Fig. 3.11, which has a 73° phase deficit. a) the first two seconds of the input signal shown in Fig. 3.12, b) correction signal, c) output signal. Because of the large phase deficit, the correction signal is delayed with respect to the input signal. The corrections for the values marked with simple, double arrows in Trace a) are similarly marked in Trace b). The correction for a peak is delayed so much that it arrives at a time when the input signal has a valley. Thus, rather than suppressing the input, the correction signal enhances it instead, which leads to unstable behavior, shown in the lower trace.

From the above discussion, it would appear that, in order to ensure closed loop stability, the phase lag of $L(j\omega)$ should be less than 180° when $|L(j\omega)| = 1$. These intuitive considerations thus lead us to the

Nyquist stability criterion. It is worth noting that, at frequencies where the gain is higher than unity the phase lag can be substantially higher than 180° without leading to system instability.

Chapter 4

EXAMPLES

This section illustrates many of the concepts and issues relevant to FCS design by using optical tracking of aircraft and laser frequency noise suppression as examples. Various aspects of the examples will be used in later sections to clarify details of the design process.

1. CAMERA FOR AIRCRAFT TRACKING

The example of a camera with the capability of automatically tracking an aircraft is used to introduce the concepts of **sensor range** and **lock acquisition**, and to discuss some basic issues related to the use of motors and gears.

1.1. Camera: Range and Lock Acquisition

In order for a tracking system to work, the loop shown in Fig. 2.1 needs to be closed, which in turn requires that all elements in the loop function properly. In the example shown in Fig. 4.1, the CCD camera, which is the sensor in the system, can provide an error signal only if the aircraft is within the field of view. If this is not the case, the input to the compensator cannot be processed into a signal which leads to successful tracking. In other words, the variable to be controlled, in this example the angle between the line of sight to the aircraft and the optical axis of the camera, needs to be within what is called the range of the sensor, in this case the field of view of the CCD camera. When the aircraft is outside the camera's field of view, the tracking cannot work and the system is free-running. If, on the other hand, the image of the aircraft is on the CCD, the camera can track the plane. The latter situation is described by saying that the system is locked, or in-lock. Thus the need arises to ensure a transition, called lock acquisition, between the free-running and the locked states of the system. The arrangement of Fig. 2.1 does not explicitly contain any provision for lock acquisition. What is needed is some sort of **search algorithm**, which eventually brings the system within the sensor range. In some fortunate cases one can rely on the random nature of the disturbance variable $x_d(t)$ to bring the system within sensor range; otherwise, an explicit search mechanism has to be designed into the system.

1.2. Electrical Motors as Actuators

Fig. 4.2 shows in some detail the means used to adjust camera pointing. The main features of the pointing arrangement are:

- A mount which provides the camera with two tilt degrees of freedom, azimuth and elevation.

- Two electrical motors that are used to point the camera. Because electrical motors are typically designed to turn at rather high rates, reduction gears, sometimes included in the motor package, are needed for smooth pointing of the camera in very small angular increments.

While motors are simple and frequently encountered devices, their use in a closed-loop system requires some care. This is especially true in the context of linear systems which form the object of this book. The fact to keep in mind is that motors, in particular when used with reduction gears, are affected by friction. The simplest way to describe an electric motor is to assume that it is driven by a current source-type amplifier, such that the current in the windings of the motor is $i(t) = k_1 v(t)$, where $v(t)$ is the voltage at the input to the amplifier and k_1 is a constant. Since the torque at the shaft of the motor is proportional to the current, $\Theta(t) = k_2 i(t)$, one can write:

$$\Theta(t) = kv(t) \tag{4.1}$$

Examples

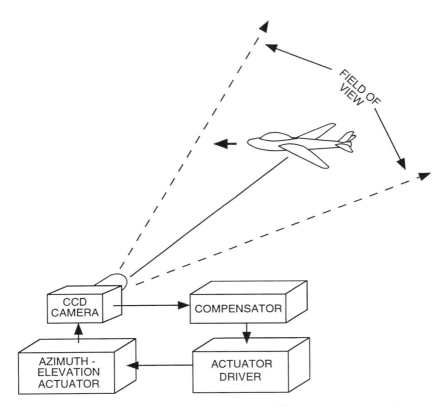

Figure 4.1. Example of a tracking system: a CCD camera is required to keep the image of the aircraft at the center of the CCD. In order to accomplish this, the camera itself in conjunction with appropriate software are providing an error signal when the image is off-center. The error signal is processed by the compensator and fed to an azimuth-elevation actuator, which corrects the pointing of the camera to keep the image centered.

which is a remarkably simple relationship. The pitfall with this actuator is illustrated in Fig. 4.3a. The real job required of the motor is to rotate the camera. However, due to friction in the motor and the gear, a minimum torque needs to be applied before the shaft breaks loose and starts moving. This introduces a strong nonlinearity in the relationship

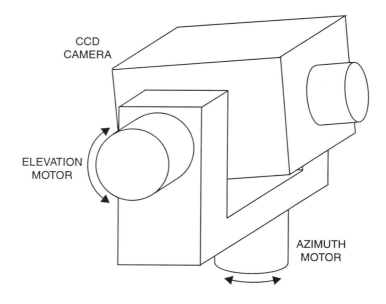

Figure 4.2. Azimuth-elevation pointing arrangement for a CCD camera. The camera is set up on a fork mount and is pointed using motors with reduction gears.

between shaft velocity and the voltage at the amplifier input. This nonlinearity is a serious problem when it comes to make the closed-loop system work. As shown in fig. 4.3b, the nonlinearity caused by friction can be substantially reduced by using an amplifier that behaves as a voltage source. This works if the amplifier can sustain large currents and if the resistance of the motor windings is small. When the motor is still, even a small voltage at the amplifier output will cause a large current to flow through the motor windings, and the correspondingly large torque will get the shaft to move. As the motor turns faster and faster, the back EMF reduces the current in the winding, providing a natural torque regulator and resulting in an almost linear velocity/voltage relationship. The residual nonlinearity is caused by the finite resistance of the windings and the by the nonzero output impedance of the amplifier.

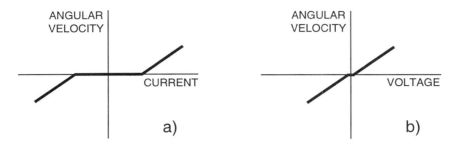

Figure 4.3. Effect of current/voltage drive for a motor. a) Current drive results in constant torque at the motor shaft, which leads to a significant nonlinearity due to friction in the motor and the gear. b) Voltage drive leads to a marked reduction of the nonlinearity associated with friction.

2. ND:YAG LASER FREQUENCY STABILIZATION

Lasers are the optical analog of electronic oscillators. Most lasers include an **optical resonator**, which allows the optical field to build up at frequencies at- or close to resonance. This is the equivalent of the tank circuit in an electronic oscillator. Field amplitude decay as a result of losses is prevented by adding an **amplifying medium**, or **gain medium** to the resonator. When energy is supplied to the gain medium, lasing, i.e. a sustained emission of a coherent light beam is the result. The operation of any gain medium is based on an effect called **stimulated emission**, which relies on the coherent emission of photons by a system which can sustain a population inversion between two energy levels. This example discusses a particular type of laser, the monolithic Nd:YAG laser. The laser itself consists of a Nd:YAG crystal a few millimeters in size, which is the optical gain medium. The population inversion which leads to gain occurs between levels due to Neodymium (Nd) ions implanted in the lattice of an Yttrium Aluminum Garnet (YAG) crystal. The beam follows a closed path through the crystal, which is also the laser resonator. The laser frequency is determined by the length of the optical path through the crystal. Energy is supplied to the gain medium by a process called **optical pumping**, whereas light from one or several diode lasers emitting light at $\lambda \sim 810$ nm is absorbed by the crystal and causes the necessary population inversion.

2.1. Free Running Laser Frequency Noise

The frequency of a diode laser-pumped Nd:YAG laser fluctuates because of a variety of technical and fundamental effects. Slow frequency fluctuations result from changes in crystal temperature, while fast frequency fluctuations are caused mainly by fluctuations in the intensity of the light used for optical pumping. For the purpose of this design example, the following free-running frequency variations will be considered:

- **Drift**
  Over time scales of 100 s and longer, the laser frequency makes excursions as large as 1 GHz, due to changes in ambient temperature.

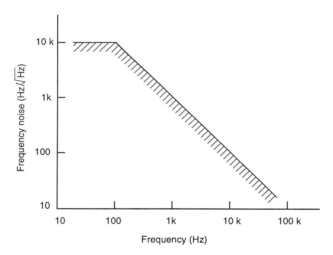

Figure 4.4. Example of upper limit for the free-running frequency noise of a ∼ 1 W monolithic Nd:YAG laser. If the plot is derived from a a frequency noise measurement, one should keep in mind that the units displayed on spectrum analyzer screen are usually root-mean-square (rms). A commonly used multiplier for converting rms units to peak-to-peak units is 5. This is based on the frequently (but not always!) encountered empirical fact that wideband noise consists of pulses with short duty cycle.

- **Noise at Intermediate Frequencies**
  The frequency noise spectrum for a monolithic Nd:YAG laser is usually a fairly complicated function of frequency. The suppression ratio

$|L(j\omega)|$ (see Eq. 2.3a on p. 14 or Fig. 2.2b on p. 14) has to be chosen such that adequate suppression is provided at the peaks in the noise spectrum. Thus, one needs to know an upper limit for the noise spectrum. For design convenience, this should be a smooth function with simple frequency dependence. A sample upper limit for the free-running frequency noise spectrum of a Nd:YAG laser is shown in shown in Fig. 4.4. The spectrum of Fig. 4.4 integrates to a total frequency error of 200 kHz rms which is equivalent to $\sim$1 MHz peak-to-peak.

- **Fast Frequency Fluctuations**
  Occasionally the laser frequency undergoes step function-like jumps up to 50 kHz over $\sim$10 ns, presumably due to effects related to the diode lasers used as a pump source. Since these jumps can be up or down, their range is $2 \times 50 = 100$ kHz peak-to-peak.

2.2. System Concept

In terms of the vocabulary used so far, x, the variable to be controlled, represents fluctuations in the frequency of the laser. The output variable $x_{fr}(t)$ is the free-running value of the laser frequency fluctuations, caused by internal and external disturbances driving the laser frequency away from a constant value. The purpose of frequency noise suppression is to reduce the frequency fluctuations. In the example discussed here, this is achieved by using a FCS which forces the frequency to track a stable reference. According to Eq. 2.3 on p. 12, $x_{fr}(t)$ is attenuated by a factor equal to the magnitude of the open loop frequency response, $|L(j\omega)|$. If the frequency noise suppression system only aims at obtaining a constant frequency, but the value of the constant frequency is not important, the corresponding FCS will be a disturbance suppression arrangement.

A commonly used arrangement for stabilizing the frequency of a Nd:YAG laser is shown in Fig. 4.5.

Sensor

Detecting laser frequency fluctuations relies on coupling the light to a rigid optical resonator, also called optical cavity, consisting of two mirrors attached to a rigid spacer, such that their optical axes coincide. This open resonator is the optical equivalent of stable radio frequency (RF) resonant systems, like cryogenic cavities or molecular absorption

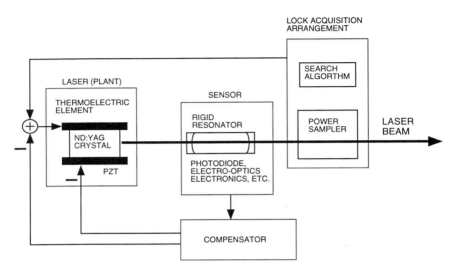

Figure 4.5. Concept of laser frequency noise suppression system. The sensor and the actuators will be discussed in this section. Lock acquisition will be addressed later.

lines, used to stabilize the frequency of RF oscillators. Typically, mirror transmission is set at a fraction of a percent, so that multiple overlapping beams propagate and interfere with each other inside the resonator. When the resonator length is an integral multiple of half of the wavelength of the light, a standing wave pattern builds up inside the resonator, a condition which is known as **resonance**. At resonance, the resonator transmission reaches its maximum value, and so does the power circulating inside the resonator. If the frequency changes by a **free spectral range, FSR**, defined as $c/2l$,[1] the cavity length becomes equal to the next higher integral multiple of the half-wavelength and another resonance is reached, as shown in Fig. 4.6a. The dependence of transmission on frequency suggests the possibility of using the resonator as a sensor for frequency fluctuations. The reference in this case is one of the resonant frequencies, which in turn is related to the length

---

[1] where c is the speed of light and l is the length of the resonator.

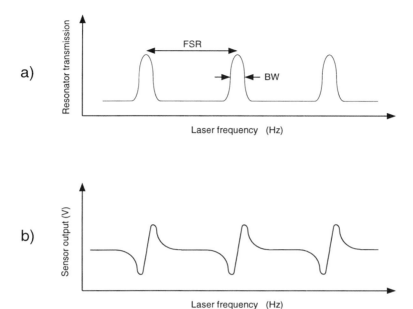

Figure 4.6. a) Light power transmission of an optical resonator versus frequency. **BW**: resonator bandwidth, **FSR**: free spectral range, defined as $c/2l$, where c is the speed of light and l is the length of the resonator. b) Main features of output signal from a Pound-Drever-Hall frequency sensing arrangement. The features on the above plots are associated with **resonances** of the optical cavity used a reference.

of the resonator; hence the choice of a rigid resonator, to ensure stable length. Measuring the transmitted power is not useful for FCS implementation, since it is symmetric with respect to resonance and therefore indicates only that the frequency is off resonance, but not which way it is off. Among many ways devised to overcome this limitation, the Pound-Drever-Hall method[2] has established itself for use in high per-

---

[2] A description of this method is highly technical and beyond the scope of this book. The interested reader is referred to: R. W. P. Drever, J. L. Hall, F. V. Kowalski, J. Hough, G. M. Ford, A. J. Munley, H. Ward *Appl. Phys. B*, **31**, 97 (1983)

formance laser frequency noise suppression systems. A sample output signal corresponding to the Pound-Drever-Hall sensing arrangement is shown in Fig. 4.6b. As long as the laser frequency is within the resonance bandwidth, the sensor output is approximately linear. The slope of the output, expressed in V/Hz, is the **gain** of the sensor. A remarkable feature of the Pound-Drever-Hall sensor output is that when the frequency moves out of the resonance bandwidth, the slope levels off and then changes sign. Thus, if this sensor is used in a FCS which is stable on-resonance, pulling the frequency off-resonance would render the system unstable, as negative feedback turns into positive feedback. Therefore the **resonance bandwidth** defines a **sensor range**. The laser frequency has to be within the sensor range for the system to function.

The trace in Fig. 4.6b is obtained when the laser frequency is scanned slowly through resonance. For faster scanning rates, the fact that the field is stored in the resonator over time scales $\sim 1/BW$ influences the measurement; the faster the scanning rate, the lower the measured gain. The scanning rate is also called the **Fourier frequency** in order to distinguish it from the actual laser frequency, which is the controlled variable in this case. If the controlled variable is not a frequency, this distinction does not have to be made. Dependence of the gain and phase shift on the Fourier frequency, for a resonator with 200 kHz bandwidth, is shown in Fig. 4.7 on p. 45. The two traces, amplitude (gain) and phase of the frequency response versus frequency constitute a Bode plot. The Bode plot of Fig. 4.7 can be seen as a model of the frequency fluctuation sensor, detailed enough for FCS design.

In summary, the sensor compares the laser light half-wavelength with the length of the resonator. When the latter is not an integral multiple of the former, a voltage proportional to the difference is generated at the sensor output.

Anything which changes the sensor output when the laser frequency does not change is defined as sensor noise or error. Examples of sensor noise/error are:

- Changes in resonator length due to ambient temperature variations.

- Changes in resonator length due to environmental vibration and acoustic excitation.

- Changes of the optical path inside the resonator due to fluctuations in air pressure, if the resonator is in air.

Examples 45

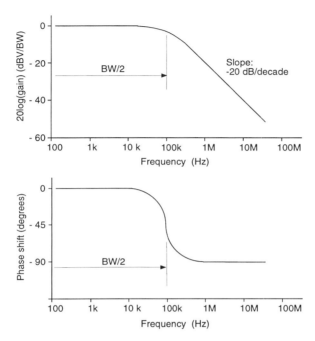

Figure 4.7. Sample Bode plot of output characteristic from a Pound-Drever-Hall frequency fluctuation sensing arrangement. Note that both the gain and the phase characteristics are flat from DC to roughly half the resonator bandwidth.

- Electronic noise associated with the sensor.

Actuators
The actuators for correcting the laser frequency are a provision for laser crystal temperature control and a piezo-electric device (PZT) bonded to the laser crystal. Both change frequency by changing the optical path inside the laser crystal. Temperature control provides for changing the frequency over a range of about 5 GHz, while the PZT can be used to tune the laser over approximately 100 MHz. Typical Bode plots for the actuators are shown in Fig. 4.8. It can be seen from the plots that the temperature control actuator has a substantially higher gain than the PZT.

The time constant typical for heat exchange between the thermoelectric device and the laser crystal is ~10 s, which is why temperature-induced frequency changes start showing a substantial phase shift and roll-off around 0.1 Hz. PZT-induced frequency changes do not show significant phase shifts up to a few hundred kilohertz. Therefore, the temperature control can be used to effect slow, but large frequency corrections, while the PZT is appropriate for small amplitude, fast frequency corrections.

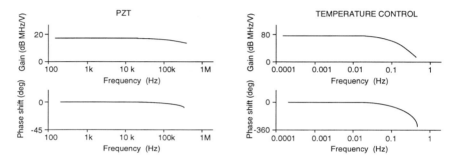

Figure 4.8. Sample Bode plots for frequency-tuning devices in a typical monolithic Nd:YAG laser made by Lightwave Electronics. The variable for the horizontal axes is the Fourier frequency.

2.3. Tracking Requirement

A tracking requirement can be formulated by requiring the laser frequency to follow a command signal. More often however, it is required that the laser frequency be insensitive to environmental factors, like changes in ambient temperature or the presence of vibration. This is called a **frequency stability requirement**[3] and could be formulated as follows:

1 The laser frequency shall be within a fixed 10 kHz range, peak-to-peak, once lock to the reference resonator has been acquired, i.e.

---

[3] In this context the word "stability" means lack of sensitivity to environmental changes and should not be confused with FCS stability.

Examples

once the laser frequency is within sensor range and the FCS loop is closed and functioning. A short discussion of "lock" and "lock acquisition" has been given in Section 1. Note that this is a really strict requirement. Indeed, since $\lambda = 1064$ nm corresponds to $\nu = 3 \cdot 10^{14}$ Hz, the laser frequency has to be stable to one part in $3 \cdot 10^{10}$.

2 There is no requirement as to what the center frequency of this range should be.

3 Lock acquisition should be automatic, and the acquisition sequence should be initiated automatically whenever the system is out of lock.

4 Once lock has been acquired, it should be maintained for continuous time intervals of at least 10 hours.

2.4. Environmental Parameters

It will be assumed that the laser is to be used in air, in a normal laboratory environment. Since ambient temperature changes cause changes in laser frequency and reference resonator length, temperature is probably the most important environmental parameter in this example. Thus, the requirement will be made that the laser maintain the specified frequency stability (not to be confused with FCS stability) over a temperature range of 5°C, which exceeds the normal temperature variation in an air-conditioned room.

2.5. In-Band, Out-of-Band Frequency Ranges

Comparing the requirement of Section 2.3 that the laser frequency be stable within 10 kHz peak-to-peak with the free-running error of Section 2.1, one notes the following:

1 The long term frequency drift has to be attenuated.

2 The frequency noise up to ~100 kHz has to be attenuated.

3 The frequency "jumps" over 10 ns time scales need to be attenuated.

In practice, it is exceedingly difficult to operate a closed loop system at ~ 100 MHz. Thus, the FCS for this problem will most likely be designed to attenuate frequency noise at Fourier frequencies up to ~ 100 kHz, which would take care of Points 1,2 above, while the frequency jumps of Point 3 will have to be addressed by other means, e. g. filtering by passing the entire laser beam through the reference resonator. As a consequence, it is convenient to divide frequencies (in this case Fourier frequencies) into two categories:

- **In-Band (IB)**
  Frequencies where tracking errors exist **and** are attenuated by the FCS. For the present example, frequencies from DC to 100 kHz are IB.

- **Out-of Band (OB)**
  Other frequencies. For the present example, frequencies over 100 kHz are OB.

It is important to realize that even though OB frequencies are not addressed by the FCS, they are relevant to the design problem:

1. OB disturbances can prevent achievement of the tracking requirement, as in the present example.

2. OB disturbances can disrupt FCS operation, if they drive the parameter which has to be controlled outside the range of the sensor.

3. In the presence of nonlinearities, OB effects can be frequency shifted IB. For example, steady high frequency OB signals could be rectified by slow electronic components. The FCS wold perceive this as a frequency deviation and attempt to correct it. If this effect is large enough to push the laser frequency outside the range of the sensor, the loop will be broken and the FCS will cease to function.

II

DESIGN AND IMPLEMENTATION

Chapter 5

DESIGN PRINCIPLES

This Chapter presents a general discussion of the FCS design principles used in this book.

1. DESIGN APPROACH

Designing a FCS comes down to specifying the sensor, the compensator and the actuator(s) such that proper reference tracking and disturbance suppression are achieved. An implied requirement is that the FCS be oscillation free, i.e. stable, as closed loop operation can produce instability. Laying out a strategy for designing a FCS with specified tracking accuracy and stable operation is the objective of this Chapter. The procedure described is not unique or general, nor does it achieve optimum performance in a rigorous sense. Nonetheless, it provides a practical approach to many FCS design problems encountered in the laboratory. If all design goals cannot be met because of too many constraints, priority should be given to solving the problems which would prevent the system from working altogether. For example, even if disturbance rejection is not as good as desired, the system should at least be stable. The main aspects which need to be addressed by FCS design are summarized in the Box below, their ordering reflecting the "system-should-at-least-work" requirement.

> **Main FCS Design Topics**
>
> - System stability
> - Range and bandwidth of actuators
> - Magnitude of open-loop frequency response for required performance
> - Noise

1.1. Assumptions

The FCS design approach followed here is based on the following assumptions, which reflect the reality of the laboratory:

1. The end product of the design process is a system which complies with pre-determined performance requirements. Performance is not required to be optimal in a rigorous sense.

2. The starting point for the design consists of:

    - A concept for the FCS
    - A tracking performance requirement
    - Some data on the free-running behavior of the variable $x(t)$
    - Some knowledge of the environment in which the system is required to perform

3. Noise contributed by the compensator, $N_G(f)$, can be made negligible with respect to reference noise (or error) $N_r(f)$ by proper compensator design. Noise contributed by the actuators and their drivers is negligible compared to $N_G(f)$.

4. The design process has to contend with inaccurate performance requirements and incomplete knowledge of the spectrum $X_{fr}(f)$ of the free-running values of $x(t)$, the variable to be controlled.

Design Principles 53

5 The dynamics of components such as sensors and actuators are only partially known.

6 The behavior of electronic components outside the linear regime is unknown or difficult to incorporate into a reasonably simple model.

7 Specifying subsystems and components, prototyping and experimental performance assessment are essential to the design process.

1.2. Error Budget

The contributions to the error affecting the controlled output x_{Cl} of the system are read off Eq. 2.3, p. 12:

1 Sensor noise $N_s(f)$.

2 Reference error, $N_r(f)$, which consists of a slow variation of the nominal reference value, $N_{r,slow}(f)$ due e. g. to thermal expansion, creep, component aging, drift in the electronics and other slow effects, and a fast fluctuation $N_{r,fast}(f)$ due to fast environmental and internal fluctuations, e. g. acoustic excitation, vibrations, and electronics noise.

3 Compensator noise $N_G(f)$ can usually be made negligibly small by adequate design. Therefore, this term should not contribute to the noise budget.

4 Residual output fluctuation with spectrum $X_{fr}(f)/L(j\omega)$, due to finite open-loop frequency response.

In a well designed system output noise (or error) thus consists of sensor and reference errors and residual output fluctuations. In practice, it is usually easier to:

- achieve high $|L(j\omega)|$ at low frequencies, since $L(j\omega)$ has to roll off at high frequencies in order to ensure system stability, and

- to obtain low $N_s(f)$ and $N_r(f)$ at high frequencies, because suppressing drift and other low-frequency sensor and reference errors is generally difficult.

The noise budget is therefore expected to be dominated by sensor and reference errors at low frequencies and by residual output fluctuations at high frequencies.

1.3. Design Sequence

A summary of the FCS design process is presented in this Section and will be discussed in detail in Chapter 6. The design flow is represented schematically in Fig. 5.1.

One notable characteristic of this process is that it includes designing the plant, as well as selecting the **system architecture**. The latter consists of choosing the sensors and actuators and of deciding on their relative placement with respect to each other and the plant, as well as the choice of a lock acquisition arrangement, if such is needed. This is in contrast with the common view on FCS design, where the plant, the sensors and the actuators as well as their relative placement are given, and the design focuses exclusively on the compensator. The former approach is obviously less constrained and therefore likely to lead to better system performance with a lower level of effort and expenditure. In this top-down approach, one starts with defining the system and one proceeds with selecting the architecture. Only then does one move on to design the compensator. The design cycle is completed by building and testing the system. The required performance is achieved by iteration, including changes in the plant and in the system architecture if necessary. Below is an outline of the level-by-level, step-by-step design cycle.

1 **System Level Considerations**

   Typically, a FCS is added to a passive, meaning no-closed-loop system, in order to improve its performance in some specific way, for example by reducing its sensitivity to external conditions. This assumes that there is a specific performance requirement which the passive system does not meet. Since the addition of the FCS comes at a price in terms of cost, complexity and reliability, it is wise to consider first the possibility of redesigning the passive system in order to meet the performance requirement without the use of feedback control. If it turns out that this is not possible, it is desirable to use the "uniform distribution of pain" rule:

   (a) Push the design of the passive system to the point of diminishing returns.

Design Principles

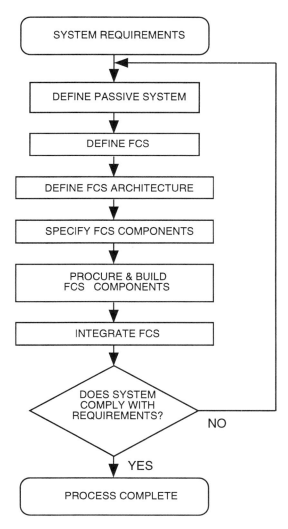

Figure 5.1. Schematic representation of the FCS design process. The significance of this diagram resides in the fact that it includes all aspects of the overall system design.

(b) Design the simplest FCS capable of bridging the performance gap.

Note: from this perspective, achieving optimum performance for the FCS is quite irrelevant; achieving the system performance specification at a minimum cost is what matters.

Redesign of the passive system should target two aspects:

- Improvement of passive system performance, e. g. reduced sensitivity to changes in ambient temperature.

- Simplifying passive system (plant) dynamics, that is obtaining a plant with the simplest possible frequency response. This is important because the plant response is part of the overall open-loop frequency response, thus a simpler plant will lead to a simpler open-loop frequency response and therefore to a simpler compensator. Chapter 11 explains in detail how plant resonances can lead to closed loop instability.

2 **Control System Architecture**

Before compensator design can begin, a FCS **architecture** needs to be selected. This will include:

(a) Number and type of sensors.

(b) Number and type of actuators.

(c) Location of sensors and actuators, within the overall system and with respect to each other. In particular, sensor/actuator collocation should be pursued vigorously. This will reduce signal propagation delays in the loop and the associated phase lags, and will therefore lead to improved system stability with a simpler compensator.

(d) Deciding which parts of the FCS will be analog and which ones will be digital.

(e) Establishing if there is a need for an explicit lock acquisition provision.

3 **FCS Design Steps**

(a) Sensor specification
- Range
- Noise
- Gain

Design Principles 57

- Linearity within range
(b) Actuator specification
(c) Open loop frequency response specification
(d) Compensator specification
(e) Provisions for lock acquisition
(f) System integration, i.e. making the FCS work
    - Achieving stable closed loop operation
    - Measuring the actual open loop gain
    - Checking if sensor and actuator ranges are sufficient
    - Measuring FCS performance
    - Measuring lock acquisition efficiency.
(g) Iterating the design to achieve the desired performance with some margin.

The meaning of some of the above guidelines will become apparent when the issues are discussed in detail in the following Chapter.

The fact that iterating the design is explicitly made part of the process removes the need to go into excruciating detail in the initial phase of the work. This makes it possible to move forward fast and settle most details gradually, as test data become available.

2. INPUT DATA FOR FCS DESIGN

The input to the design process consists of the following information:

- A concept for the system, including the type of sensor and actuator(s) and the lock acquisition method.

- A requirement on tracking performance which indicates the maximum allowed departure of the controlled variable x_{Cl} from the reference value e_r, i.e.

$$|X_{Cl}(f) - E_r(f)/B(j\omega)| \leq \mathcal{T}(f) \qquad (5.1)$$

It is useful to find out, by interaction with the customer, how much safety margin has already been included in the requirement, as a function of frequency. Knowing the frequencies where the requirement is "hard" and those where it is just a goal or a "desirement" can result in a big difference in design effort and cost.

The tracking requirement should also include a statement on the minimum time over which the system is expected to track continuously.

- Some knowledge of the disturbance x_d of the parameter of interest, preferably its spectrum $X_d(f)$. This information is likely to contain a large initial uncertainty, which will be reduced when measurements with the loop closed become possible.

- The range for relevant environmental parameters under which the system is required to display the specified tracking performance. It is important to identify the environmental parameters which will have the heaviest impact on system performance. Temperature is often among them.

Chapter 6

DESIGN AND TROUBLESHOOTING

The steps of the design process, outlined in the previous Chapter, are explained in detail in what follows. The points where one has to chose between moving on or iterating some previous steps are highlighted. The design activity is considered complete when the new system has been tested and found to perform as specified. The structure of the FCS design flow has been discussed in Section 1.3 of the previous Chapter. Here it will be assumed that:

1. The top-level requirements for the system have been defined. In other words, it has been determined what the system is supposed to accomplish.

2. The frequency response of the passive plant has been simplified as much as it seemed possible with a reasonable level of effort[1] and the need to supplement it with a FCS has been established.

The remainder of the FCS design steps are discussed in the following sections as follows:

- System architecture related topics

    1. Sensor Specification, Section 1
    2. Actuator specification Section 2

---

[1] Chapter 11 discusses ways to keep the frequency response of the plant simple, for the case of flexible mechanical systems.

3 Lock acquisition arrangement Section 5

- Open-loop gain specification and compensator design Sections 3,4
- Building, testing and refining the system until the specified performance has been achieved Sections 6,7,8.

1. SENSOR SPECIFICATION

1.1. Sensor Range

Several considerations which help choosing the sensor range $\mathcal{R}_s$ will be discussed below. They are all related to the circumstance that, when the variable $x(t)$ is outside the range of the sensor, the loop is open and the FCS is not operational.

1 IB Requirement
When the system performs according to specification, the variable $x(t)$ still shows a residual error with respect to the reference. An upper bound to the IB peak-to-peak residual error is calculated by integrating the spectrum of the tracking requirement defined in Eq. 5.1

$$e_{IB} = 5 \left[\int_{IB} \mathcal{T}^2(f) df \right]^{\frac{1}{2}} \quad (6.1)$$

If the system is to maintain lock, the residual error should not be capable of pulling $x(t)$ outside the sensor range:

$$2e_{IB} < \mathcal{R}_s \quad (6.2)$$

2 OB Requirement
Like in the IB case, the range of the sensor has to accommodate the OB error, that is

$$2e_{OB} < \mathcal{R}_s \quad (6.3)$$

3 Lock Acquisition Considerations
Before lock is acquired, $x(t)$ is typically outside the range of the

Design and Troubleshooting 61

sensor. The loop is therefore open, which usually means that at least some of the amplifiers making up the compensator are saturated, as explained in the box below.

> According to Eq. 2.4 on p. 12, the correction signal is approximately equal to the free-running output $x_{fr}(t)$ when the loop is closed, regardless of the value of the compensator gain, as long as the loop gain is larger than unity. If the amplifier chain is properly designed, none of the amplifiers in the chain are saturated. If the loop is open, however, the error signal is amplified by the compensator according to the available gain. If this gain is high, some amplifiers in the chain are likely to saturate, in particular the stages closer to the actuator.

When the lock acquisition subsystem brings $x(t)$ within the range of the sensor, closed loop operation sets in, and tracking starts to take effect. However, due to amplifier saturation, attenuation of the tracking error is less than described by Eq. 2.3. The system needs some time to come out of saturation and accomplish the nominal error attenuation. This can happen only if $x(t)$ is within sensor range even when the residual error is larger than under nominal tracking conditions. In other words, the range of the sensor has to be wider than required by the IB condition above.

> A useful starting point, chosen somewhat arbitrarily, is to require that sensor range be the larger of $20e_{IB}$, $2e_{OB}$.

If during testing it turns out that lock acquisition is difficult, one may have to redesign the compensator electronics for faster recovery from saturation. This is usually simpler, easier and faster than respecifying and rebuilding the sensor.

Going back to the example of the laser, $e_{IB} = 10$ kHz, and $e_{OB} = 100$ kHz peak-to-peak (see p. 41). Thus, the range of the sensor, which is

1.2. Sensor Error

According to Eq. 2.3, p. 12, the tracking error $X_{Cl} - E_r/B$ can be no lower than the sensor error N_s. Therefore, sensor error is bounded by the tracking requirement:

$$|N_s(f)| < |\mathcal{T}(f)| \tag{6.4}$$

for all IB frequencies.

For design purposes, it is convenient to split sensor error and reference error into slow (e. g. drift, creep, etc.) and fast (arbitrarily defined as $f > 0.1$ Hz) components:

$$\begin{aligned} N_s &= N_{s;slow} + N_{s;fast} \\ E_r &= E_{r;slow} + E_{r;fast} \end{aligned} \tag{6.5}$$

A good electronics design will ensure that sensor error dominates at all IB frequencies:

$$\begin{aligned} |E_{r;slow}| &\ll |BN_{s;slow}| \\ |E_{r;fast}| &\ll |BN_{s;fast}| \end{aligned} \tag{6.6}$$

While designing the sensor for satisfactory noise performance, it is worth keeping in mind the following considerations:

1. The sensor is usually long lead. The success of the design may depend to a large extent on identification and careful analysis of the slow and fast contributions to sensor error and ensuring they are low enough to comply with Eq. 6.4.

2. In terms of the ultimate performance of the FCS, $BN_{s;slow}$ and $E_{r;slow}$ are equivalent. However, they affect tracking performance by different mechanisms:

    - In the absence of any other error, and under stable FCS operation, $x_{Cl}(t)$ tracks the reference; if the reference drifts, $x_{Cl}(t)$ drifts with it.

Design and Troubleshooting 63

- In the absence of any other error, $x_{Cl}(t)$ will be offset from the reference by the inverse Laplace transform of BN_s. If N_s becomes larger than the sensor half-range, the FCS will loose lock. Thus one should ensure successful closed-loop operation by selecting the electronic components and construction techniques for the sensor such that $\left[\int |N_s(f)|^2 df\right]^{1/2}$ is kept well below the range of the sensor.

Referring to the example of laser frequency noise suppression, Points 1,2 above translate as follows:

1 Since the resonant frequency changes with resonator length as $\Delta \nu/\nu = \Delta l/l$, and the stability requirement is that $\Delta \nu/\nu < 3 \cdot 10^{-11}$, the length of the reference resonator has to satisfy $\Delta l/l < 3 \cdot 10^{-11}$. If thermal expansion is the only cause of length change, one needs to ensure that $\alpha_t \Delta t < 3 \cdot 10^{-11}$, where α_t is the coefficient of thermal expansion. With a resonator made of ULE (ultra-low expansion glass) for which $\alpha_t \sim 10^{-9}/°C$, resonator temperature variations have to be limited to $\Delta t < 0.03°C$, when the ambient temperature may change over a 5°C range. Clearly, both thermal insulation and temperature control of the reference resonator will be necessary.

Likely contributions to $N_{s;fast}$ are acoustic and seismic excitations of the resonator spacer. While it is hard to make a general statement regarding the level of these disturbances, it is very likely that, in order to meet the frequency stability requirement that $\Delta \nu/\nu < 3 \cdot 10^{-11}$, one will have to place the reference resonator on a seismic isolator, in vacuum.

2 A typical range for the sensor output is $\sim \pm 1$V peak-to-peak for the sensor bandwidth of 200 kHz, as illustrated by Fig. 4.7 on p. 45. Thus sensor electronics noise should contribute significantly less than ± 0.1 V at the sensor output in order to allow the system to stay within range once lock has been acquired. Meeting the requirement that the laser frequency fluctuate less than ± 10 kHz (see the tracking requirement for the example of the laser, in Section 2.3, p. 46) means keeping electronic contributions to the sensor output below ± 0.05 V. Likely electronic contributions to sensor error are :

- Temperature-dependent offset voltages at the input of the op-amps
- Pick-up at multiples of the line frequency

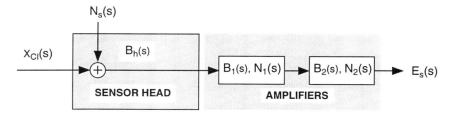

Figure 6.1. Multiple-stage sensor configuration. A distinction is being made between the sensor head, where the parameter to be measured is converted into electrical signals, and the amplifiers (two stages are shown), which may include some filtering. The overall sensor transfer function, measured in units of V/[units of x], is $B(s) = B_h(s)B_1(s)B_2(s)$. Each stage is characterized by a transfer function and a noise/error contribution.

- Electronics noise and electromagnetic interference not related to the power line

All these effects are hard to assess quantitatively in advance. In the early phase of the design it is enough to make sure that the various terms listed above do not add up to more than ± 0.1 V. This will allow the loop to be closed and FCS tests to be carried out, while the sensor itself is subjected to independent tests aimed at reducing its output error to less than ± 0.05 V, consistent with the requirement on the limit to frequency fluctuations.

1.3. Sensor Transfer Function

A more detailed block diagram for the sensor is shown in Fig. 6.1. The sensor head is defined as the part which converts the parameter $x(t)$ into an electrical signal. The purpose of a subsequent stage(s) as shown in Fig. 6.1 is:

- To amplify the signal from the sensor head in order to make sensor error the dominant error in the system, in particular over compen-

Design and Troubleshooting

sator noise/error. This is another way of saying that high sensor gain $|B(j\omega)|$ reduces the total noise affecting $x_{Cl}(t)$ (Eq. 2.3, p. 12).

- To buffer sensor head output from the compensator input, e. g. in case of a piezo-electric sensor with ~ 1 MΩ output impedance and a compensator with a low-noise input stage with ~ 1 kΩ input impedance.

- To allow the implementation of filtering needed for proper operation of the compensator and of the actuator. For example, it may happen that high frequency noise or pick-up are showing at the output of the sensor head, but cannot be tolerated by the slow compensator electronics, where it may get rectified and thus generate an unwanted DC offset in the system. One solution is a filter/amplifier stage after the sensor front end.

The signal at the sensor output is:

$$E_s(s) = B(s)\left(X_{Cl}(s) + N_s(s)\right) + B_1(s)B_2(s)N_1(s) + B_2(s)N_2(s) \quad (6.7)$$

where $B(s) = B_h(s)B_1(s)B_2(s)$. A good sensor design will ensure that the only significant contribution to sensor output error will come from the sensor head itself, that is:

$$\begin{aligned} |B_h(j\omega)N_s(f)| &\gg |N_1(f)| \\ |B_1(j\omega)N_1(f)| &\gg |N_2(f)| \end{aligned} \quad (6.8)$$

where $\omega = 2\pi f$. For the example of the laser stabilization arrangement, the tracking requirement corresponds to 50 mV at the sensor output. This is high enough to dominate compensator noise almost regardless of compensator design, thus no sensor output amplification is necessary. The only issue than is to design the sensor front end such that its own error/noise is much lower. e. g. by a factor 10, than the tracking requirement.

1.4. Sensor Nonlinearity

Sensor nonlinearity, defined as x-dependence of the sensor gain, which is the derivative of sensor output with respect to x, is not a crucial issue as long as it is small. Several examples of large nonlinearities, with their possible adverse effects on FCS operation, and tentative remedies are listed below.

- The signal at the input of the sensor falls outside the range of the sensor, that is $|x_{fr}(t)|_{max} > \mathcal{R}_s$ when the loop is open or $|x_{Cl}(t)|_{max} > \mathcal{R}_s$ when the loop is closed. This is a condition where the gain of the sensor drops to zero and the loop cannot function. The remedy is to provide a lock acquisition function in order to bring the variable $x_{fr}(t)$ within sensor range when the loop is open, and to increase loop gain in order to keep $x_{Cl}(t)$ within range once the loop is closed.

- Gain reduction due to nonlinearity, e. g. saturation, can be so severe that it may prevent the FCS to ensure proper tracking. The remedy is to design enough excess gain into the system to keep the overall gain high enough for proper tracking even when nonlinearity sets in.

- Signal distortion may cause phase shifts which erode the phase margin enough to render the closed-loop system unstable. The remedy is to design sufficient phase margin into the system to accommodate phase shift due to nonlinearity.

- A change of sign of the sensor gain would render the system intrinsically unstable. There is no easy remedy for this situation.

A general but costly way to mitigate the existence of nonlinearities is to define the sensor range as the interval of x where the system is at worst mildly nonlinear, and then design the FCS to operate within this limited range. This entails increasing the reach of the lock acquisition arrangement, and a further increase in the magnitude of the open-loop gain $|L(j\omega)|$. It may happen that keeping the system inside the linear range defines a tougher requirement than the original tracking specification.

The laser frequency noise suppression problem presents an example of extreme nonlinearity. As it can be seen from Trace b) in Fig. 4.6 (p. 43), the signal decays rapidly outside the bandwidth of the reference resonator, rendering the closed-loop system unable to track. The range of the sensor is thus determined by the resonance bandwidth, and a provision for lock acquisition, as shown in Fig. 4.5 on p. 42, has to be included in the design.

2. ACTUATOR SPECIFICATION

Actuators are the part of the FCS which correct tracking errors detected by the sensor and filtered and amplified by the compensator. In order to ensure proper tracking, the actuators, which under the definitions used here include the actual correction device and its driver, have to satisfy two conditions:

1. According to Eq. 2.3 on p. 12 the effect of the FCS is to cancel the disturbance $x_d(t)$ and replace it with the reference value $e_r(t)/B$. Therefore, the correction range of the actuators has to be wide enough to cover the maximum in-band variation of $x_d(t)$, or, equivalently, $x_{fr}(t)$:

$$\mathcal{R}_{act} > 5 \left[\int_{IB} |X_{fr}(f) - E_r(f)/B(j\omega)|^2 df \right]^{\frac{1}{2}} \qquad (6.9)$$

2. The actuator has to be "fast" enough to be able to correct for errors at the high end of the IB range. In more concrete terms, this means that the gain of the actuator should not fall off significantly over the in-band frequency range and that the phase shift due to the actuator should not be to large around the intended unity gain frequency of the FCS.

Returning to the example of the laser stabilization system, the requirement is to cancel frequency fluctuations of up to 1 GHz. The temperature control, which is capable of tuning the laser frequency over ~5 GHz, has plenty of range. However, inspection of the corresponding part of Fig. 4.8 shows that for frequency fluctuation rates, i.e. Fourier frequencies above 0.1 Hz temperature control just cannot correct the frequency of the laser anymore. Temperature control is to "slow" an actuator for correcting frequency fluctuations at Fourier frequencies from ~0.1 Hz to ~100 kHz, as required. A "faster" actuator is needed; as the upper box in Fig. 4.8 shows, the PZT has the necessary speed, as its magnitude plot hardly decreases at all up to 100 kHz. On the other hand, the PZT has a correction range of only 100 kHz, and is not able to cover the entire range of expected frequency error. This situation, calling for a slow, wide range actuator to be used in conjunction with a faster, narrower range one, is fairly common. The art of designing such arrangements for close-to-maximum gain and stable operation is covered in Chapter 7.

3. SHAPING THE OPEN-LOOP RESPONSE

The open-loop frequency response has to be specified while keeping in mind two criteria :

- The need to provide sufficient gain for achieving the prescribed tracking performance.

- The frequency dependence of the open-loop gain has to be consistent with closed-loop stability.

According to Eq. 2.3 (p. 12), if noise is disregarded, $|X_{Cl} - E_r/B| \simeq |X_{fr}/L|$. On the other hand, the design has to ensure that the tracking requirement of Eq. 5.1 on p. 57, $\mathcal{T}(f) \geq |X_{Cl}(f) - E_r(f)/B(j\omega)|$, is satisfied. From these two equations it follows that:

$$|L(j\omega)| \geq \frac{X_{fr}(f)}{\mathcal{T}(f)} \qquad (6.10)$$

which defines the minimum magnitude of the open loop frequency response $L(j\omega)$, as a function of frequency, necessary in order to satisfy the tracking requirement.

The derivation of a lower bound for the open loop gain L is illustrated for the tracking requirement for the laser frequency noise suppression system of Section 2.3, Chapter 4. The requirement is that the laser frequency be within 10 kHz of the reference (assumed to be more stable than this requirement), while the low-frequency drift is up to 1 GHz, and the frequency noise at higher Fourier frequencies is bounded by the plot of Fig. 4.4, p. 40.

The somewhat arbitrary assumptions will be made that the open loop gain shall be chosen so that the residual low-frequency and high-frequency errors are equal, each approximately equal to 3 kHz, and that the residual noise above 10 Hz be white, e. g. constant as a function of Fourier frequency.

Design and Troubleshooting 69

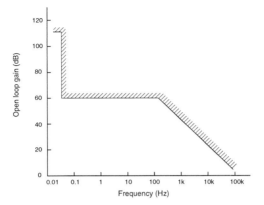

Figure 6.2. Example of lower bound of open-loop gain for laser frequency noise suppression.

Real-life design work often demands that somewhat arbitrary choices be made, in order to keep the process moving. It is important that these choices be documented so that they can be tracked and modified, in case the design is faced with excessive difficulty. This is one of the circumstances which make design iteration necessary.

The low-frequency gain should then be at least 1GHz/3 kHz= 300000 = 110 dB. A total noise level of 3 kHz over a frequency span of 100 kHz corresponds to a constant spectral density of 10 Hz/Hz$^{1/2}$. Thus, from Fig. 4.4 the gain has to be at least 1000 = 60 dB up to 100 Hz, with a $1/f$ roll-off at higher frequencies. The result of this process is illustrated in Fig. 6.2. The immediate consequence of using Eq. 6.10 is a vertical slope at low frequencies. It is impractical to realize this kind of steep frequency dependence for the gain. Therefore, one has to modify the frequency dependence to a shape which is practically feasible, while encompassing the requirement set by Eq. 6.10.

At the frequency where $|L(j\omega)| = 1$ the slope of the gain plot in Fig. 6.2 is 20 dB/decade. This corresponds to a phase lag of 90° (Appendix A), which, according to the Nyquist criterion predicts stable closed-loop behavior. In general, however, the lower bound for $|L(j\omega)|$, obtained from Eq. 6.10 may not satisfy the Nyquist criterion (Chapter 3). If this is the case, $L(j\omega)$ should be chosen such that the following criteria are satisfied:

FEEDBACK CONTROL SYSTEMS

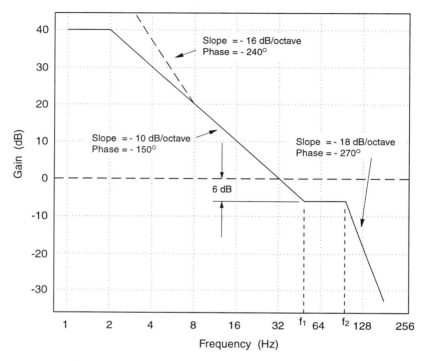

Figure 6.3. Example of "optimum" open loop gain. The solid line decreases towards the unity gain frequency at the maximum slope, -10 dB/octave, consistent with an adequate phase margin of 30°. The -18 dB/octave slope, chosen for the higher frequency range, causes a 270° phase lag, a fraction of which is felt at lower frequencies and may thus degrade the phase margin. The gain step between f_1 and f_2, called **Bode step**, prevents the higher phase lag corresponding to the steeper slope to "spill" towards the unity gain frequency and degrade the phase margin. The dotted line shows how the gain can be increased towards lower frequencies.

- $|L(j\omega)|$ is higher than the lower bound derived using Eq. 6.10.
- The phase margin is at least 30°.

Fig. 6.3 shows an example of a frequency response $|L(j\omega)|$ that satisfies the Nyquist stability criterion and has just about the highest slope from the unity gain frequency towards lower frequencies, consistent with a phase margin of 30°. Before going into a more detailed discussion of Fig. 6.3, it is worth noting that unlike previous Bode diagrams in this book, here the frequency axis is divided into octaves, rather than decades, in order to provide adequate resolution for quick eyeballing of $L(j\omega)$. A sharp slope has been chosen above f_2 in order to filter out noise at frequencies where the loop has little effect because the open-loop gain is less than one. The flat portion between f_1 and f_2 is called the **Bode step**. Its purpose is to maintain the phase margin against some additional phase lag kicked towards lower frequencies by the portion of the plot above f_2, which has high phase lag because of its steep slope.

In some cases, the -10 dB/octave slope does not lead to a sufficiently high gain at lower frequency as required by the minimum gain specification. In that case, one should consider enhancing the slope by - 6 dB/octave starting about two octaves below the unity gain frequency, as shown with the dotted line in fig. 6.3. The choice of a two-octave spacing between the ramp-up point and the unity gain frequency results in a reduction of the phase margin by $\sim 10°$. This is roughly compensated by the presence of the Bode step. If even more gain is needed at lower frequencies, an additional increase in slope by -6 dB should be applied starting at four octaves below the unity gain frequency. A practical way to accomplish these -6 dB/octave increases in slope is given on p. 88.

While 30° of phase margin are adequate for disturbance rejection, it leaves room for possibly intolerable levels of ringing and overshoot in systems which need to respond to step function-like commands. As shown in Fig. 3.4 on p. 23, 45° of phase margin lead to a relatively smooth response to a step function command. 90° of phase margin would actually ensure critically damped response, but a design with such a high phase margin would suffer from reduced gain slope, and thus reduced gain at lower frequencies, meaning reduced performance. The solution is to keep the phase margin at 30° and pass the command through a **prefilter** before applying it to the closed-loop system. The prefilter in this case would be a notch filter which attenuates the command in a narrow frequency band around the unity gain of the FCS. This removes from the spectrum of the command signal the portion which excites system ringing, while letting most of the energy in the command to enter the FCS. For a detailed discussion of prefilters and their use, see the book by Lurie and Enright quoted earlier in this book.

Finally, it is worth noting that the specification for $L(j\omega)$ is more easily used in subsequent FCS design steps if it is expressed in terms of a DC gain, and the number and positions of poles and zeros.

4. COMPENSATOR SPECIFICATION

Throughout this book the compensator is seen as a piece of electronic hardware whose function is to contribute a frequency response G which together with the response functions of the sensor and of actuator makes up an open loop frequency response L that achieves the specified FCS performance. Thus, compensator specification consists of a frequency response specification and a hardware-type specification.

4.1. Compensator Frequency Response Specification

The compensator transfer function G is obtained from the definition of the open loop gain L given in Chapter 2, p. 11:

$$G(s) = \frac{L(s)}{B(s)H(s)A(s)P(s)} \qquad (6.11)$$

where B, A, H and P are the transfer functions of the sensor, the actuator, the actuator driver and the plant, respectively.

4.2. Compensator Hardware Specification

Once the compensator transfer function has been derived using Eq. 6.11, the stage is set to generate a complete specification which can be handed over to an electronics design engineer who will work out a circuit diagram. The specification also has to provide the necessary demands and constraints placed on the packaging of the circuitry. The circuit diagram and the packaging requirements are then passed on to the fabrication shop where the compensator transfer function will be translated into a piece of hardware.

In order to facilitate electronics design and construction, and in order to enable trouble-shooting for proper close-loop operation, the compensator specification should include :

1 The transfer function as calculated with Eq. 6.11, expressed as a DC gain and a list of poles and zeros.

2 A normal signal input which should be characterized by:

- Range of variation of the input signal.

Design and Troubleshooting 73

- Input impedance, which should be high enough to prevent overloading the sensor output.
- Input referred offset, which should be less than sensor offset at the output of the sensor.
- Upper bound to input referred noise spectrum, in $V/\sqrt{Hz}$, which should be less than the noise at the output of the sensor.
- Type of connector or other means of interfacing with the sensor, e. g. a radio link or a digital interface (bus).

3 A test input which allows injection of a test signal under operating conditions. This input can be implemented as a summing point at the input of the compensator, or at the input of any of the compensator amplifier stages.

4 A normal output which should be characterized by:

- Output range, selected to approximately match the range of the actuator(s).
- Output impedance, which should normally be much lower than the input impedance of the actuator driver(s) which follow downstream.
- Maximum current drive requirement, determined by the input impedance of the following actuator driver and the output range.
- Type of connector or other means of interfacing with the actuator driver.

5 A test output for each amplifier stage, to assist the testing process. This could be a test point on the printed circuit board. A more convenient implementation is as a connector on the front panel of the compensator, which would make it possible to observe the signal using an oscilloscope, without the need to open the box.

6 A front-panel switch for changing the sign of the signal. This switch will help ensure that negative feedback is established once the loop is closed.

7 A front-panel gain adjustment knob, for changing the compensator gain, and with it the open-lop gain L, by two decades above and two decades below the nominal gain. This provision is necessary because the initial design is rarely accurate enough to predict the exact value of the gain "as built," and also because the necessary

gain is somewhat uncertain, given the uncertainty in the initial design data.

8. Input and output protection against shorting and overload. This provision is necessary because closed-loop operation is likely to engender system oscillation at one time or another, in particular during the trouble-shooting phase. Conditions may then occur which could damage the compensator or the actuator. For example, if the actuator is a piezo-electric device and is driven directly by the compensator, oscillation may result in voltages much higher than the maximum rating of both the actuator and the compensator, which may thus incur irreparable damage. A voltage clamp would provide adequate protection in this case.

9. The cut-off frequency and roll-off rate of a low-pass filter at the output of the compensator. This filter is often necessary in order to prevent noise above the unity gain frequency f_0 to propagate around the system and possibly cause saturation and thus incapacitate the loop. The design of this filter should ensure that:

    - The reduction of high-frequency noise is adequate.
    - The cut-off frequency $f_{\text{cut-off}} > f_0$ and the roll-off rate are such that the phase shift caused by this filter at the unity gain frequency f_0 does not reduce the phase margin to unacceptably low levels. The implementation of this filter is related to the steep part of $|L(j\omega)|$ at frequencies above f_2 in Fig. 6.3.

These two requirements are obviously contradictory[2] and reconciling them may sometimes add a significant burden to the FCS design.

Comments on Amplifier Design

Once the trade-off between performance and stability has been resolved by choosing a given value for the phase margin, it is essential that this value be maintained within a few percent throughout the lifetime of the compensator, against the combined forces of component aging and changing environmental parameters, especially temperature. The components which determine the frequency response of the compensator should be resistors and capacitors, as they can be obtained with good long term and temperature stability at relatively low cost. Opamps, on

---

[2] see Appendix A for the relationship between filter roll-off and phase shift.

Design and Troubleshooting

the other hand, have characteristics which can be quite different from those listed in the data sheets, as variations by factors of 2 are not uncommon even within the same production run. The discussion of feedback amplifiers on p. 163 then leads to the concern that intrinsic bandwidth limitations of the amplifiers and the associated phase lags may degrade the phase margin to a point where the system could show unacceptable ringing or even become unstable. Worse yet, if the design is marginal, changes in phase lag caused by variations in ambient temperature may cause an intermittent instability which is very hard to troubleshoot. With the considerations listed on p. 163 in mind, the following approach is recommended:

- Choose opamps with the highest possible gain-bandwidth product, while satisfying the noise and range requirements of the design.

- Assume that the worst case for gain-bandwidth, listed in the data sheets under "minimum," applies.

- Make sure that any frequency below $1.5 f_0$ the gain of the amplifier stage is at least a factor 100 below the open-loop gain of the opamp used for that stage.[3]

- If the above condition cannot be met with a single amplifier stage because of high open-loop gain requirements, replace the single amplifier with a chain of amplifiers, such that the gain of each individual stage is well below the open-loop gain of the corresponding opamp at all relevant frequencies.

5. LOCK ACQUISITION

If the system is initially outside the range of the sensor, the loop cannot be closed, and the FCS is unable to perform its tracking function. While $x_{fr}(t)$ is outside the range of the sensor, it is likely that the amplifying stages of the sensor and compensator will be saturated, thus driving the actuator toward the edge of its range. There are two distinct lock acquisition regimes:

---

[3] f_0 is the unity gain frequency of the FCS.

1 **Spontaneous Lock Acquisition**
 This occurs if the natural fluctuations of $x_{fr}(t)$ are large enough to bring the system within sensor range occasionally. If the variation of $x_{fr}(t)$ is slow enough to give the amplifiers time to come out of saturation while $x_{fr}(t)$ dwells within the range of the sensor, lock acquisition will "happen" without any special addition to the FCS. If spontaneous locking appears to be a possibility, the electronics should be designed with quick desaturation, that is fast amplifiers in mind. Choosing fast amplifiers is a particularly important consideration if $x_{fr}(t)$ undergoes fast fluctuations and the range of the sensor is narrow. This is the case when, for example, the frequency of a laser is to be locked to a narrow-band optical resonator or to a narrow atomic line.

2 **Induced Lock Acquisition**
 If $x_{fr}(t)$ would "never" (meaning very seldom) enter sensor range, explicit provisions need to be added to the FCS. The remainder of this subsection contains a short discussion of induced lock acquisition.

Locking can be induced in many different ways. Fig. 6.4 presents one example which illustrates what's involved. The main components of the arrangement in Fig. 6.4 are:

- An "out-of-range" detector which continuously analyzes the output of the FCS sensor for anomalous behavior indicating that the variable $x_{fr}(t)$ is outside the sensor range, which means that the system is out of lock.

- A set of switches.

- A search pattern generator.

- A **HOLD** circuit.

Lock acquisition proceeds as follows:

1 Absence or loss of lock is detected by the out-of-lock detector.

2 The locking sequence is initiated when the out-of-lock detector sets both switches to the "**U**" (unlocked) position. SW2 thus disconnects the actuator from the compensator, which is saturated anyway, while SW1 connects the search pattern generator to the actuator input.

Design and Troubleshooting 77

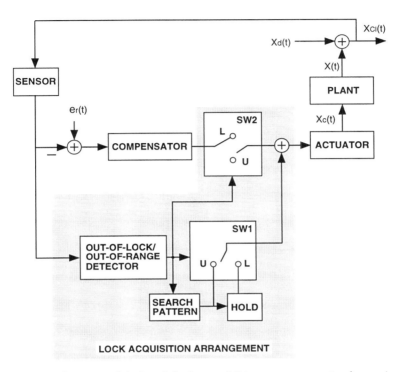

Figure 6.4. Concept of induced lock acquisition arrangement, shown in the shaded area of the picture. **SW1,2**: switches; **U**: system in unlocked (out-of-range) state; **L**: system locked.

3. The search pattern generator is triggered, which causes the actuator to scan x_{fr}.

4. When x_{fr} moves within sensor range,[4] the FCS amplifiers in the sensor and the compensator come out of saturation.

5. The out-of-range detector senses that the system is now within range. Accordingly, it sets both switches to the "**L**" (locked) position, and

---

[4]Strictly speaking, x_{fr} is being replaced with x_{fr}+the effect of the actuator on the plant output.

commands the hold circuit to continuously apply the last value of the search pattern generator to the actuator input. The FCS loop is thus closed and the system operates as designed, while the lock acquisition arrangement is basically out of the loop. Lock has been acquired.

Example 1: Laser Frequency Noise Suppression
For the stabilized laser shown in Fig. 4.5, the in-range state corresponds to the laser field resonating with the reference resonator. In this state, the maximum amount of light is transmitted through the resonator. The out-of-lock/out-of-range detector can thus be implemented as a transmitted power measurement combined with a threshold detector; an in-range state is declared when the transmitted light exceeds a preset threshold. The search pattern could be a succession of step increases in laser crystal temperature, effected by the thermoelectric element. As a result, the laser frequency would change in steps, until it comes close to a reference cavity resonance. At that point, searching is stopped and the loop closed.

Example 2: Aircraft Tracking
For the tracking camera shown in Fig. 4.1, the in-range state corresponds to the image of the aircraft being present on the CCD array. The out-of-lock/out-of-range detector can thus be implemented as an algorithm which recognizes the shape or the motion pattern of the plane. The search pattern could be an outward spiraling motion of the camera, executed using the tip-tilt actuator. When the image of the plane is detected, the search is stopped and the loop is closed, so that the camera can track.

6. SYSTEM INTEGRATION: MAKING IT ALL WORK

After designing the FCS and having all its parts physically built, the next phase is to put the components together and achieve the desired performance. This stage in the project, also known as **system integration**, is carried out mainly in the laboratory. In order to perform the various tests and measurements described in what follows, it is useful to have at least some, but preferably all of the following instruments:

Design and Troubleshooting 79

1. One or several oscilloscopes. Since a common form of FCS malfunction is related to amplifier oscillation at a frequency much higher than the design value of the unity gain frequency f_0, it is desirable to have at least one oscilloscope fast enough to visualize oscillation in the fastest amplifier in the system.

2. A function generator, for creating the test signals needed to assess the health of various parts of the electronics.

3. A data acquisition system capable of recording transient behavior of signals of interest.

4. A network analyzer, which is needed for measuring frequency response functions of amplifier/filter stages in the compensator. Typically, a network analyzer consists of an internal signal generator, and two input channels. As shown in Fig. 6.5, a frequency response measurement proceeds as follows:

   - A sine-wave signal generated by the internal source is applied to both Channel 1 and the input of the system under test.
   - For each frequency in a pre-defined range, the machine calculates the frequency response, which is the ratio between the output and the input to the device under test, i.e. the ratio:
   (Channel 2)/(Channel 1)
   - The magnitude and the phase of the (Channel 2)/(Channel 1) ratio are displayed on the network analyzer screen and stored in digital form.

5. A spectrum analyzer, sometimes called dynamic signal analyzer. This instrument is useful in measuring the noise at various points in the loop. This is important for assessing the tracking performance of the system and for evaluating the performance of individual amplifiers. Often the spectrum analyzer and the network analyzer are implemented in the same versatile instrument.

6.1. Achieving Closed-Loop Operation

Of all the steps in the design and testing of a FCS, this one is without doubt the most challenging. Indeed, in most cases proper closed-loop

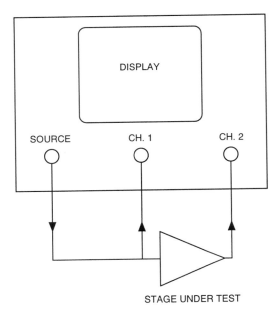

Figure 6.5. Use of a network analyzer for measuring the frequency response of an amplifier stage.

operation will not be established when a new system is being put together for the first time, even though the individual parts of the system, i.e. the sensor, the compensator and the actuator have been tested and are in good working order. The presence of tracking or the lack thereof can be determined by observing the signals e_s and e_c at the sensor and actuator output test points shown in Fig. 6.6. When the switch **SW** in Fig. 6.6 is open and the system is thus free-running, there should be "life" at the sensor output $e_s(t)$, while $e_c(t)$ should be a flat line.[5] When the switch **SW** is closed, the system should track. If it does, the sensor signal should be attenuated, while the correction signal changes with time. For systems with very high open-loop gain, the sensor output may appear "frozen" during closed-loop operation, due to the large

---

[5]See Eqs. 2.4,2.5 on p. 13.

Design and Troubleshooting

degree of disturbance suppression, according to Eq. 2.3. Therefore, as summarized in Table 6.1, when the loop is closed by throwing the switch **SW**, one should see the action move from $e_s(t)$ to $e_c(t)$. This assumes that $x_{Cl}(t)$ is closely tracking the reference $e_r(t)$. On the other hand, if reference tracking does not take place, some of $e_{s,G_1,G_2,G_3,c}(t)$ may be at their extreme (saturation) values. In this case, the transfer of action from $e_s(t)$ to $e_c(t)$ when **SW** is closed will not happen. If $x_{fr}(t)$ is completely out of sensor range, $e_s(t) = 0$ and there will be no change in any of the test signals when **SW** is closed, except perhaps for some changes in DC levels. These considerations are summarized in Table 6.2.

Loop Condition	Sensor Output	Correction signal
Open	Lively	Flat
Closed	Flat	Lively

Table 6.1. Appearance of sensor output e_s and actuator driver output e_C for open- and closed-loop conditions.

Change in appearance of e_s, e_C	Diagnostic
e_s goes flat, e_C goes lively	Closed-loop OK
e_s stays lively, e_C stays flat	Saturated amp
e_s stays flat, e_C stays flat	Sensor out of range

Table 6.2. Summary of loop condition diagnostic by changes in appearance of sensor output e_s and actuator driver output e_C, when switch **SW** is closed.

Since the various operating parameters are vastly different in the closed-loop and open-loop regimes, testing the system open-loop in order to find why the loop can't be successfully closed is going to be only marginally useful. The troubleshooting approach discussed in what follows is based on the observation that, when the system is brought within sensor range, all the components of the system are in a linear regime for a short time, until linearity is lost because the loop fails to acquire or maintain lock. Specifics of the transition from linearity to severe non-

82   FEEDBACK CONTROL SYSTEMS

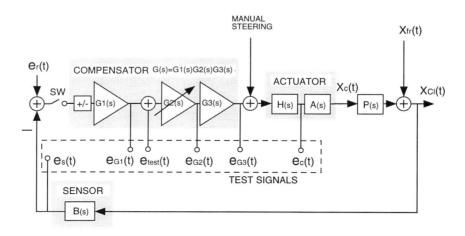

Figure 6.6. Modified version of Fig. 2.1, used in the discussion on troubleshooting the FCS and testing its performance. Since much of the testing is carried out by looking at signals using an oscilloscope, the figure emphasizes the time-domain aspect of the signals. **SW**: switch for selecting open- or closed-loop operation; $\boxed{+/-}$: sign switch for ensuring negative feedback; e_s: signal at sensor output; $e_{G1,2,3}$: signals at outputs of various compensator amplifier stages; e_c: correction signal; e_{test}: signal at test input. The second stage of the compensator has variable gain for adjusting the overall open-loop gain. The input labeled "MANUAL STEERING" is used to manually bring the system within the range of the sensor in the evaluation phase.

linearity, e. g. which amplifier outputs are getting stuck at one of the power supply voltages, hold the clues needed to make the loop work. Troubleshooting can thus be conducted along the following three-step procedure:

1. Bring the system within sensor range manually, by using the input normally used for lock acquisition, as shown in Fig. 6.6.

2. Record the linear-to-nonlinear transition at one or several of the output test points $e_s(t)$, $e_{G1,2,3}(t)$, $e_c(t)$ shown in Fig. 6.6.

Design and Troubleshooting 83

3 Analyze the time record.

The tests will require an oscilloscope and a sufficiently fast transient recorder.

The most likely causes of closed-loop failure are:

1 Closed-loop instability

2 Insufficient open loop gain

3 Insufficient actuator range

A short discussion of each of these follows.

Closed Loop Instability

If the loop is closed and the system is steered manually, observing the sensor output e_s will indicate when the system comes within range; the sensor output shows variation with time, within the known range for sensor output voltages. If lock is not acquired at this point because of closed loop instability, the sensor output will show an oscillation which builds up to a point where some amplifier in the system saturates, as shown in the example of Fig. 6.7. A first quick diagnostic consists of changing the sign of the open-loop gain by using the sign switch built into the compensator. If the cause of instability was positive feedback, it should now be possible to close the loop. If the loop appears to be working, changing sign should cause instability as in Fig. 6.7.

If changing the sign does not help, there is a genuine closed-loop instability. In other words, the phase lag for $|L(j\omega)| = 1$ is more than 180°. To further investigate this possibility, one should attempt to set the sign correctly. Many times, this is not practically possible, and all the tests will have to be carried out for both signs. In some cases, the frequency of the oscillation is visibly different for the two signs settings. The correct sign then probably corresponds to the switch setting where the oscillation frequency is close to the nominal unity gain frequency of the FCS.

Assuming that the feedback sign is correctly set, the next thing to look for is an incorrect overall gain value. The example shown in Fig. 6.8 illustrates one such situation. The phase lag at 300 Hz is $\sim 40°$ less than 180°. Thus, if the unity gain is at 300 Hz, as in the middle trace, the closed loop should be stable. For both the upper and lower traces, however, unity gain occurs at frequencies where the phase lag is higher than

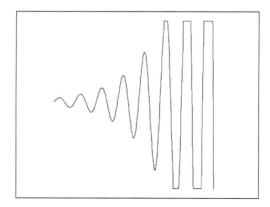

Figure 6.7. Build-up of oscillation as a result of closed loop instability, as seen at the sensor output. Saturation is visible at the right hand side of the trace. Since the loop is closed, it is not obvious from the figure where in the system saturation occurs. To locate the first amplifier to saturate, one has to look at the output test points downstream of the sensor.

180°, which will cause closed-loop instability. Sometimes the specific design for $L(j\omega)$ is such that instability can occur only when the gain is too high. In any event, if improper gain is the cause for instability, both the diagnostic and the remedy consist of changing the overall gain by using the manual gain control included in the design of the compensator.

If both changing the sign and changing the overall gain fail to eliminate the instability, the designer is likely to have one of the following problems:

1 Unaccounted for dynamics in the system cause additional phase lag.

2 Compensator as built introduces more phase lag at unity gain than allowed for in the design.

In either case one will have to perform a piece-by-piece measurement of the open-loop frequency response $L(j\omega)$, followed by appropriate redesign. When measuring the frequency responses of the various parts of the loop, keep in mind the following:

Design and Troubleshooting

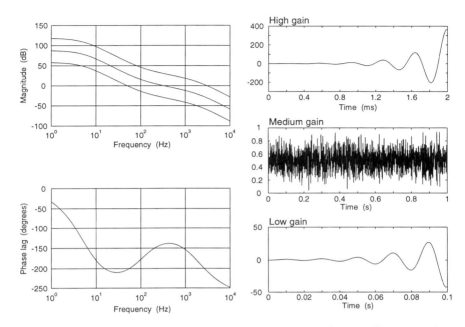

Figure 6.8. Example of system that goes unstable when the overall gain is either too high or too low. The system is driven with random noise. On the left are Bode diagrams of a system for three overall gain settings. On the right are simulations of the closed loop system output for the three gain settings. In each case, the only input to the system is a signal taking random values between 0 and 1. For the high gain setting, unity gain occurs at 3 kHz, where the phase lag is approximately 200°. As expected, the simulation shows rapid instability buildup, with a strong component at approximately 3 kHz, the unity gain frequency. For the low gain setting, unity gain occurs at 50 Hz, where the phase lag is just over 200°. This system is also expected to oscillate, and indeed the simulation shows an oscillation building up at approximately 50 Hz. For the intermediate gain setting, unity gain occurs at 300 Hz, where the phase lag is $\sim 140°$. This setting satisfies the Nyquist stability criterion and is thus expected to "behave." The simulation shows that indeed it is stable, with the output being equal to the random signal applied to the input.

- If the compensator has very large gain, it may be saturated when the loop is open, due to small electronic offsets amplified by the high gain up to or beyond the power supply rail. If this is the case, the frequency response of the compensator needs to be measured stage-by-stage using the intermediate test points, and the individual responses multiplied to yield the overall response.

- In all piece-by-piece and stage-by-stage measurements, one needs to make sure that the input- and output loadings of the pieces or stages match the conditions prevailing within the integrated system.

- The frequency response of the actuator and of its action on the plant is harder to measure if the variable x is not a voltage. Specialized setups may be necessary for these measurements.

- The above comment also applies to the sensor.

Insufficient Open Loop Gain

If the system is free of instability, lock acquisition will occur once either the disturbance $x_d(t)$ or the manual input have steered the system output within sensor range. Disregarding noise, according to Eq. 2.5 on p. 13 the sensor output will be:

$$E_s(f) = \frac{B(j\omega)}{L(j\omega)} X_{fr}(f) \tag{6.12}$$

If $|L(j\omega)|$ is not high enough, $e_{s;p.t.p.} = 5 \left[\int |E_s(f)|^2 df\right]^{1/2}$ might exceed the range of the sensor for large values of $x_{fr}(t)$. If this is the case, the scope trace corresponding to the sensor output will behave as shown in Fig. 6.9. Then, loss of tracking occurs when the sensor signal reaches its lower or upper limit. The immediate thing to try is to increase the overall loop gain by using the manual gain control of the compensator. This remedy is limited by:

- The limited range of any gain adjustment.

- Increasing the overall gain moves the unity gain frequency to higher frequencies, so that the phase margin becomes lower and lower, until the system becomes unstable.

If this simple remedy does not help, one will have to measure $X_d(f)$, the spectrum of $x_d(t)$, while the system is tracking, and increase the gain selectively around frequencies where $X_d(f)$ peaks. Often one finds that the gain needs to be increased at lower frequencies. The obvious thought is to increase overall gain and add a pole in order to take out

Design and Troubleshooting 87

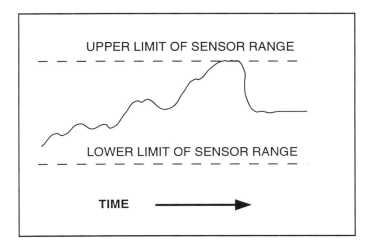

Figure 6.9. Time record of sensor output for a situation in which lock is lost due to insufficient open-loop gain. When the range of the sensor is exceeded, the sensor output goes to zero and the loop is effectively open.

the gain increase at higher frequencies. This wouldn't work, because the pole introduces 90° of phase shift above the pole frequency, which most likely would cause the system to oscillate. What one should try instead is to use the **lag-lead** circuit of Fig. 6.10, which adds a pole and a zero to the transfer function. This arrangement provides an increase in gain below the frequency of the zero, as desired. The roll-off ceases approximately at the position of the zero, which also cancels the phase lag. If the zero is at least one decade below the unity gain frequency of the FCS, stability will be affected only marginally by this addition to the compensator. If the open-loop gain has been provided with a Bode step, as shown in Fig. 6.3 on p. 70, the zero can be pushed to within two octaves from the unity gain frequency, which results in more gain increase at lower frequencies.

Insufficient Actuator Range
If tracking for extended time intervals fails even though the system is

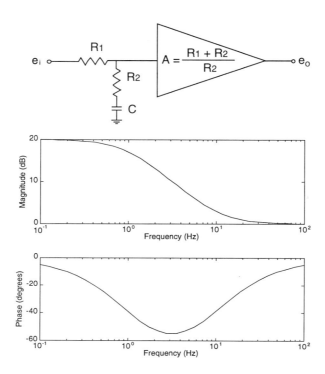

Figure 6.10. Lag-lead circuit used to increase open-loop gain at low frequencies. The circuit introduces a pole $f_P = 1/2\pi(R_1+R_2)C$ and a zero at $f_Z = 1/2\pi R_2 C$. With the amplifier as shown, the gain below the pole is $(R_1+R_2)/R_2$, while the gain above the zero is 1. The Bode plot shows an example where $f_Z/f_P = 10$, with the pole at 1 Hz and the zero at 10 Hz. At 100 Hz, the residual phase lag is only a few degrees.

stable and there is enough gain to keep the signal within the range of the sensor, the prime suspect is a limitation in the range of the actuator. Indeed, according to Eq. 2.3 on p. 12 the correction signal $x_c(t)$ is approximately equal to the free-running variable $x_{fr}(t)$, when L is high and noise is disregarded. Therefore, with reference to Fig. 6.6, p. 82 the actuator driver output is:

$$E_c(f) = \frac{X_c(f)}{A(j\omega)} \qquad (6.13)$$

Design and Troubleshooting

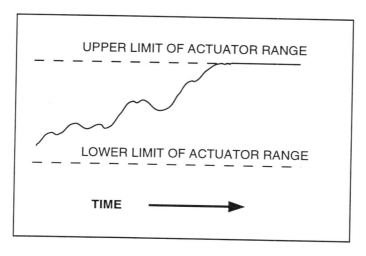

Figure 6.11. Time record of actuator driver output $e_c(t)$ for a situation where tracking fails as a result of insufficient actuator range.

If the free-running variable becomes too large, the output of the actuator driver will fail to increase enough for proper tracking, i.e. its final stage will saturate, and tracking will be lost. An example of saturation at the driver output is shown in Fig. 6.11. In many cases, $e_c(t)$ will retain its saturated value even after $e_s(t)$ becomes zero when $x_d(t)$ leaves the range of the sensor because of tracking failure, because of the high gain in the system which amplifies small DC offsets beyond the output range of the actuator driver.

6.2. Measuring the Open-Loop Frequency Response

Since the open-loop gain L is crucial in achieving the specified system performance, it is important to determine what its actual magnitude and phase versus frequency are and compare them with the design values. With reference to Fig. 6.6, (p. 82) ideally one would open the switch

SW, inject a sine-wave at the input of G_1, measure the corresponding sensor output and take their ratio, by using a network analyzer. Varying the frequency of the sine-wave over the band of interest would yield the open-loop frequency response. This direct approach is impractical in many cases, mainly because:

- With the loop open (i.e. with the switch **SW** open), the free-running output may be outside the range of the sensor, and thus no useful signal would be present at the sensor output.

- The combined gain of the compensator and actuator driver may be large enough to force the actuator driver output into quasi-permanent saturation, whereas no change will be enforced on the plant and thus no signal would be present at the sensor output.

Neither of the two points above applies if the loop is closed, which is possible only when there is a useful signal at the sensor output and if the actuator driver is not saturated.

An open-loop gain measurement with the loop closed is described in what follows. A sine-wave $e_\text{test} = \alpha \sin(2\pi f t)$ is injected at the test input of G_2, the corresponding output $e_{G2} = \beta \sin(2\pi f t + \varphi)$ is measured and their ratio $\rho(f)$ is calculated as a function of frequency. Note that $\rho(f)$ is a complex quantity which describes the amplitude and the phase of the ratio between the output and the input signals. When the loop is open, ρ is:

$$\rho_\text{open}(f) = \left[\frac{e_{G2}}{e_\text{test}}\right]_\text{open-loop} = G_2(j\omega) \qquad (6.14)$$

where $\omega = 2\pi f$. The calculation of the corresponding ratio for closed-loop operation, $\rho_\text{closed}(f)$, is similar to the derivation of Eq. 2.1 (p. 11), yielding:

$$\rho_\text{closed}(f) = \left[\frac{e_{G2}}{e_\text{test}}\right]_\text{closed-loop} = \frac{G_2}{1+L} + \frac{AG_1G_2}{1+L} \cdot \frac{x_{fr}}{e_\text{test}} \sim \frac{G_2}{1+L} \qquad (6.15)$$

where all upper-case variable are functions of $j\omega$, $L = BG_1G_2G_3HAP$, and the x_{fr} term was neglected assuming that e_test is large. $L(j\omega)$ is obtained by comparing Eqs. 6.14,6.15:

$$L(j\omega) = \frac{\rho_\text{open}(f)}{\rho_\text{closed}(f)} - 1 \qquad (6.16)$$

6.3. Measuring the Free-Running Variable

At the beginning of the design process, the spectrum or other quantitative descriptions of the free-running variable are usually poorly known. Since the tracking performance of the system relies on $x_{fr}(t)$ suppression by the loop gain, which is set by the designer to a certain magnitude, it is important to make a reliable assessment of the size of $x_{fr}(t)$.

Using Eq. 2.4 for high values of $|L(j\omega)|$ and for $E_r(f) = 0$ the spectrum $X_{fr}(f)$ of the free-running variable can be calculated if the spectrum of the input to the actuator driver, $E_c(f)$, is known:

$$X_{fr}(f) = -P(j\omega)A(j\omega)E_c(f) \tag{6.17}$$

$P(j\omega)$ and $A(j\omega)$ are known from the design process and from laboratory tests of the system, and $E_c(f)$ is measured at the output of the actuator driver shown in Fig. 6.6, using a spectrum analyzer. The peak value of the free-running variable is calculated as:

$$x_{fr;peak} = 2.5 x_{fr;rms} = 2.5 \left[\int_0^\infty |X_{fr}(f)|^2 df \right]^{\frac{1}{2}} \tag{6.18}$$

6.4. Evaluating Tracking Performance

As indicated in Section 2, (p. 57), the tracking requirement consists of two components:

- Maximum departure from reference.

- Tacking robustness, measured by the minimum amount of time the system is expected to track continuously.

Maximum departure from reference refers to the quality of tracking. In some rare cases it is easy to see if tracking is appropriate. For the example where a camera is required to track an aircraft, the system performs adequately if the image of the flying plane is kept within the field of view of the camera, which is easily ascertained. In most cases however, testing for adequate tracking tends to be a more elaborate process. The maximum value of the tracking error, Δ_{peak}, is defined as:

$$\Delta_{peak} \stackrel{def}{=} \left| x_{Cl} - \frac{e_r}{B} \right|_{max} \qquad (6.19)$$

Δ_{peak} is evaluated in two steps. First, Eq. 2.3, p. 12 is used to replace x_{Cl} in the above equation. Taking also into account the fact that the noise terms add in quadrature since they are uncorrelated yields the spectrum $\Delta(f)$ of the tracking error:

$$\Delta(f) = \left[\frac{|N_r(f)|^2 + |N_G(f)|^2}{|B(j\omega)|^2} + \frac{|X_{fr}(f)|^2}{|L(j\omega)|^2} + |N_s(f)|^2 \right]^{\frac{1}{2}} \qquad (6.20)$$

Then, Δ_{peak} is calculated by integrating the power spectrum $|\Delta(f)|^2$:

$$\Delta_{peak} = 2.5 \Delta_{rms} = 2.5 \left[\int_{In-band} |\Delta(f)|^2 df \right]^{\frac{1}{2}} \qquad (6.21)$$

$\Delta(f)$ can be compared directly with the tracking requirement $\mathcal{T}(f)$, if the latter is given as a tracking error spectrum, as in Eq 5.1, p. 57. If the tracking requirement is given as a peak value, it should be compared to Δ_{peak}. If the tracking requirement is given as root-mean-square value, it should be compared to Δ_{rms}.

In principle, the integral in Eq. 6.21 should be evaluated for all frequencies. The above equation only characterizes tracking over the frequency range where the FCS is effective, that is the in-band range. This is appropriate in the context of this book, where we are interested in how good a job the FCS does in ensuring tracking.

While Eq. 6.20 gives a clear description of the tracking error as a function of the basic errors terms in the FCS, one still has to devise a way to actually measure the quantities $\Delta(f)$ or Δ_{peak}. The obvious thought is to use the sensor which is already present in the FCS. The quantity to be measured would be $\Delta_{s;peak} = |e_s - e_r|_{max}$. One would expect that $\Delta_{s;peak} = \bar{B}\Delta_{peak}$, where $\bar{B}$ is an appropriate average of the sensor frequency response $|B(j\omega)|$ over in-band frequencies. The danger with this approach has already been mentioned in Chapter 2.

Design and Troubleshooting

Using Eq. 2.5, p. 13 and taking into account the fact that the terms are random uncorrelated variables yields:

$$\Delta_s(f) = \left[|N_r|^2 + |N_G|^2 + \frac{|BX_{fr}|^2}{|L|^2} + \frac{|BN_s|^2}{|L|^2}\right]^{\frac{1}{2}} \quad (6.22)$$

The first three terms are indeed equal to the first three terms in Eq. 6.20 multiplied by the sensor frequency response B. However, the term describing the contribution of the sensor noise is suppressed by the magnitude of the loop gain $|L(j\omega)|$ at the sensor output. If the sensor error $N_s(f)$ is **guaranteed** to be small compared with the other error terms, $\Delta(f) \sim \Delta_s(f)/B(j\omega)$. In this case, since $B(j\omega)$ is known, tracking performance can be assessed by simply measuring $e_s(f)$ several times, under different conditions with the loop closed, using these measurements to determine the upper limit on $E_s(f)$, then using Eq. 6.21 to evaluate Δ_{rms} or Δ_{peak}. It should be stressed however that unless a direct measurement of $N_s(f)$ is carried out, this kind of "inside-the-loop" tracking performance evaluation is dangerously unreliable, as it would have to build on the **belief** that sensor error is small. While presumably every possible effort has been made to design a sensor with low error, the purpose of testing is to verify that the result is actually as good as desired. A more reliable test protocol is strongly advocated. Because of the enormous variety of tracking applications, it is not possible to prescribe a test recipe that is satisfactory in all cases. Nevertheless, one method which often works consists of using a sensor external to the feedback loop, as shown in Fig. 6.12. This **outside-the-loop** measurement method works if the following relationship between the errors of the internal and external sensors holds:

$$|N_{s;ext}(f)| \leq |N_s(f)| \quad (6.23)$$

In this case. tracking quality is estimated by measuring $E_{s;ext}(f)$. For example, for a system designed to regulate the temperature inside an enclosure, the external sensor would be a thermometer which is more accurate than the temperature sensor used in the loop. It is important to keep in mind the fundamental difference between the in-loop and out-of-loop measurements at the sensor output:

- **In-loop**: because of the closed loop, sensor error is impressed on the output and is canceled (and therefore hidden) at the sensor output. This is why in-loop performance measurements at the sensor output are unreliable.
- **Out-of-loop**: since the external sensor is not part of a closed loop, its error is showing at its output. Sensor error in this case is a limitation on the quality of the measurement, but does not cause any confusion.

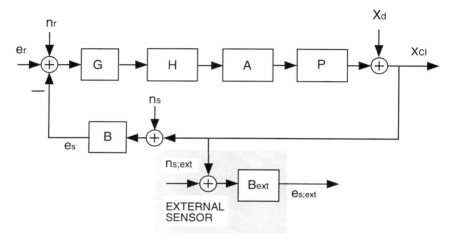

Figure 6.12. Use of an external sensor (shaded) for testing tracking performance.

Tracking performance is adequate if

$$|\Delta_{s;ext}(f)/B_{ext}(j\omega)| < \mathcal{T}(f) \qquad (6.24)$$

or

$$\left[\int_{\text{In-band}} |\Delta_{s;ext}|^2(f)/|B_{ext}(j\omega)|^2 df\right]^{\frac{1}{2}} < \mathcal{T} \qquad (6.25)$$

Tracking robustness is easy to measure by letting the system acquire lock and recording the actuator driver output $e_c(t)$. An obvious sign of

Design and Troubleshooting 95

robustness is uninterrupted tracking for at least as long as the specification requires, and preferably for times 50%-100% longer. The record of $e_c(t)$ should be compared with the output range of the actuator driver, which ideally is just a little narrower than the corresponding actuator range. In a well designed system, there is range to spare; one ad-hoc rule would be too make sure that $e_c(t)$ is within less than one third of the available range, most of the time.

7. LOCK ACQUISITION EFFICIENCY

Testing the lock acquisition arrangement consists of taking the system out of lock and letting it re-acquire a number of times. If the system re-acquires on its own within a reasonably short time every time, the design is successful. In order to make the test meaningful, it is desirable that, every time the system is taken out of lock, a different initial condition be established, so that the entire range of initial conditions likely to be encountered in practice is covered. For example, in the case of the frequency noise suppression system for the laser, every time lock is broken one would manually tune the frequency of the laser to a different point in the interval between two resonances of the reference resonator. Frequently encountered causes for difficulty in lock acquisition are:

1. **Unfavorable Initial Conditions**
   For example, in the case of the camera trying to acquire an airplane, the target search pattern may be so fast, that when the presence of the aircraft in the field of view is detected, the FCS does not have enough power, i.e. gain, to stop the search pattern before the target is lost. Possible remedies are:

   - Selection of a slower search pattern.
   - Increasing the open loop gain L.

2. **Large Offsets and/or Drift in the Sensor Electronics**
   Large errors in the electronics associated with the sensor cause the system to lock away from the center of the sensor range. According to Eq. 2.3 on p. 12, $X_{Cl} \simeq (E_r + N_r + N_G)/B - N_s + X_{fr}/L \simeq N_s$, where the last equality holds for large N_s, which is due mainly to the electronics of the sensor. If a decent design effort has been made,

the noise contribution[6] to N_s is low. Large N_s is thus usually caused by the presence of a significant offset or drift in the electronics. The above equation shows that a large N_s will push X_{Cl} off the desired value E_r/B, possibly to the very edge of the sensor range. Then, even small perturbations may be enough to push the system "over the edge" and cause it to loose lock. A large offset will cause constant lock problems. A large drift, on the other hand, may results in intermittent difficulties in acquiring and maintaining lock. Electronic drift is often caused by changes in operating temperature. The remedy is to carefully analyze the specifications of the amplifiers used in the design of the sensor electronics and select devices with sufficiently low temperature-dependent drift.

8. REFINING THE SYSTEM

After carrying out the tests described in Sections 6,7, the information needed to make the system perform to specification is available. The areas most likely to need improvement are listed below.

1 The open-loop frequency response $L(j\omega)$ may require:

- An increase in magnitude at certain frequencies, for adequate tracking and easier lock acquisition.
- An increase in phase margin, if "ringing" around the unity gain frequency is observed, and if the measured open-loop gain shows less than 30° phase margin. The connection between ringing and phase margin is addressed by Fig. 3.4 on p. 23 and the discussion preceding it.

2 If the range covered by x_{fr} is almost as wide or wider than the actuator range, the actuator selection will have to be reexamined. It is desirable that the actuator range be at least three times the range of x_{fr}.

3 In order to ensure smooth and efficient lock acquisition and a reliable lock, it may be necessary to:

---

[6]see p. 11 for a discussion of the terms **offset**, **drift**, and **noise**.

Design and Troubleshooting

- Modify the search pattern.
- Redesign the sensor electronics in order to reduce temperature-dependent drift.
- Increase the open-loop gain over certain frequency ranges.

As discussed earlier, lock acquisition is an issue only if the range of x_{fr} exceeds the range of the sensor

Once all the changes are implemented, the system needs to be put again through the tests discussed in Sections 6,7.

Chapter 7

MULTIPLE SIGNAL PATHS

In many applications, the loop configurations considered so far are insufficient for achieving the design goals. Often, adding new signal paths in parallel with the initial path can help. Three useful arrangements are presented in this section. The formal equivalence between these arrangements is emphasized, and is used to derive a stability criterion useful for designing loops with almost-maximum gain. In order to make it easier to carry out a unified discussion of control configurations with multiple paths, Fig. 2.1 is modified as shown in Fig. 7.1 on p. 100. The change consists of renaming the gain blocks, selecting a case with zero reference (control) input and rotating the various elements around the loop. These changes do not have any effect on the open-loop gain L. Therefore, reference tracking and disturbance suppression, as well as stability properties are the same for all diagrams shown in Fig. 7.1. Diagrams like the bottom one in Fig. 7.1 will be used throughout this chapter.

This chapter illustrates the fact that in many instances one can circumvent the use of full-fledged Multi-Input-Multi-Output (MIMO) formalism. All design cases covered here are of the one sensor, two actuator type. This is a special case of MIMO, called SIMO, which stands for Single-Input-Multi-Output.

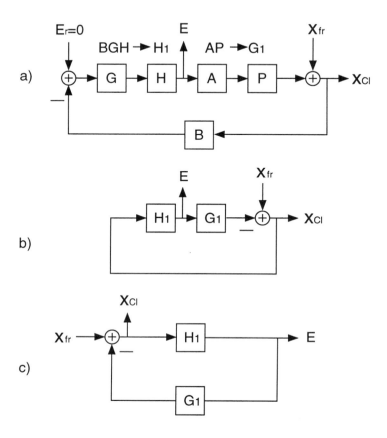

Figure 7.1. Diagram of Fig. 2.1 on p. 11 modified for the discussion of multiple control paths. a) simplified version of diagram of Fig. 2.1 with zero reference signal and zero noise. b) Same diagram with renamed blocks, $BGH \to H_1$, $AP \to G_1$. c) Equivalent diagram used in this chapter. E is a generic test signal, for checking system stability. The change from one diagram to the next keeps the open-loop transfer function $L(s)$ the same, which leaves reference tracking, disturbance suppression performance and system stability unchanged.

1. THE NEED FOR MULTIPLE PATHS: EXAMPLES

The following three examples are illustrating the use of parallel or nested paths in an FCS for solving specific control problems. Each example is derived from a real life problem. In order to make the argument easier to follow, the details have been simplified and otherwise modified to emphasize the aspects relevant to this chapter.

1.1. Parallel Actuators for Laser Frequency Noise Suppression

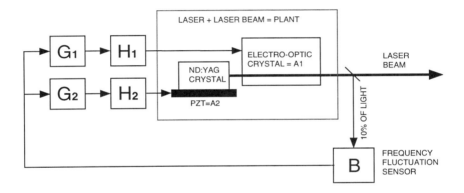

Figure 7.2. Modified version of the laser frequency noise suppression arrangement shown in Fig. 4.5 on p. 42. Elements which are not relevant to the topic of this section, the lock acquisition block and the temperature control arrangement, are not shown. The entire frequency fluctuation sensor has been condensed into a single block. A second actuator, the electro-optic crystal, has been added as a parallel control path. Further in this Section, the notations of this figure will be modified as follows for uniformity: $B \to H_1$, $G_1 H_1 A_1 P \to G_1$, and $G_2 H_2 A_2 P \to G_2$.

Applications like laser spectroscopy or high precision interferometry require the frequency of the laser light to be within a pre-determined narrow range. Commercially available lasers do generally come with substantially higher levels of frequency noise. Therefore, in order to meet the requirements of the experiment, the laser is supplemented with a subsystem which is capable of measuring the departure of the laser frequency from a certain reference, as shown in Fig. 7.2. The output from this frequency noise sensor is used in a closed loop arrangement which corrects the laser frequency. In addition to the sensor, the laser frequency noise suppression system also needs to include an appropriately designed compensator and an actuator which counteracts the frequency fluctuations by applying appropriate corrections. A very high performance laser frequency noise suppression system is described by Drever et all.[1]

As mentioned earlier in Chapter 4, Section 2.2, a common way to adjust the frequency of a Nd:YAG laser is to change the optical path covered by the light during one round trip through the laser resonator by using a piezo-electric device (PZT) bonded to the laser crystal. With the free-running frequency fluctuations as shown in Fig. 7.3 (p. 103), consider the design problem where the requirement is to reduce the frequency noise at 1 kHz and above to less than 0.001 Hz/Hz$^{1/2}$ This means that a suppression factor of 1000 is required at 1 kHz.

In order to assemble a working system, the PZT actuator needs to cover the full range of change in laser frequency, which is $\sim$ 100 MHz. For this kind of laser, a typical tuning range using the PZT is $\sim$ 500 MHz, which covers the needs of this problem with good margin. Mechanical resonances associated with the laser crystal/PZT assembly usually limit the unity gain frequency of the control loop to $\sim$ 10 kHz, which makes it practically impossible to ramp up the gain to G=1000 at 1 kHz (which is needed for reducing the frequency fluctuations 1000 times) and at the same time maintain stable closed-loop operation. This difficulty is related to the fact that the resonances may cause FCS instability. The way in which mechanical resonances cause closed loop instability is discussed in detail in Chapter 11. Since the laser comes as a sealed package, the remedies presented in Chapter 11 do not apply here. An appropriate fix consists of:

- **A frequency cut-off at the PZT drive**, below the first resonance of the PZT.

- Increasing the unity gain frequency of the loop by supplementing the system with a fast actuator. In this context, "fast" is synonymous with flat frequency response, i. e. resonance free up to frequencies

---

[1] R. W. P. Drever, J. L. Hall, F. V. Kowalski, J. Hough, G. M. Ford, A. J. Munley, H. Ward *Appl. Phys. B*, **31**, 97 (1983)

Multiple Signal Paths

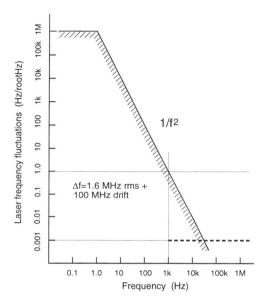

Figure 7.3. Hatched: example of upper limit to the spectrum of frequency fluctuations in a free-running laser. Dotted: upper limit to acceptable frequency fluctuations, above 1 kHz. The objective of the control problem discussed in the text is to devise an arrangement capable of suppressing frequency fluctuations by the required factor 1000 at 1 kHz.

of several megahertz. In Drever *et all*,[1] the latter is implemented by passing the laser light through an electro-optic crystal, which has the property that it changes the phase and thus the frequency of the light when a voltage is applied to electrodes coated on its side. This device has smooth frequency response up to $\sim$ 10 MHz.

The block-diagram of the resulting arrangement, shown in Fig. 7.4, consists of a common path with transfer function H_1 and two parallel paths with transfer functions G_1, G_2 corresponding to the electro-optic crystal and to the PZT, respectively. It is desirable that maximum benefit, i.e. maximum unity gain frequency, be gained by adding the electro-optic actuator. This requires that minimum propagation delay be associated with H_1G_1, which is accomplished by simplifying as much as possible the common path H_1 and using the fastest available compo-

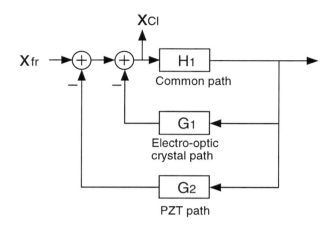

Figure 7.4. Block diagram of closed-loop system with two parallel signal paths. Note for later reference that the "fast" electro-optic path is represented by the frequency response G_1, while the "slow" PZT path is represented by G_2.

nents in the design of the hardware associated with G_1. Fig. 7.5 shows a possible choice of gain magnitudes for the two paths, with adequate gain at 1 kHz and which should be stable at the unity gain frequency of 1 MHz. The high-pass character of the electro-optic path is required in order to counter its tendency to saturate at low frequency. Following the signals along the parallel paths of Fig. 7.4 and going through the algebra one finds that the ratio between the slow/fast control efforts, i. e. between the PZT/crystal frequency correction, is approximately $|G_2/G_1|$, below the unity gain frequency f_0. The reduction of control effort in the path with narrower range is usually called **actuator desaturation**. One can see that if one path displays gain roll-off while the other one has high-pass behavior, large gain ratios and thus large desaturation factors can be achieved in this configuration.

One potential problem with this arrangement is the danger of instability at 10 kHz, where the control hand-off from one path to the other occurs. The term "hand-off" indicates unity gain ratio $|G_2/G_1| = 1$; below the hand-off frequency most of the control effort is carried by the

Multiple Signal Paths 105

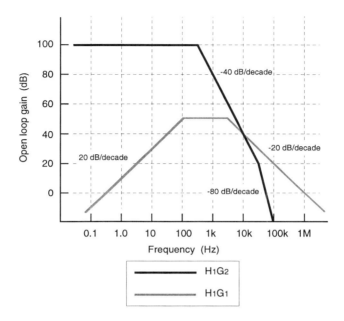

Figure 7.5. Example of possible Bode diagram for the system of Fig. 7.4.

PZT, while above the hand-off the electro-optic crystal dominates the loop. The hand-off stability issue is an important part of this discussion and will be addressed in Section 2.2.

1.2. Parallel Low Frequency Gain Boost

In certain control problems, the low-frequency gain needs to be increased. The obvious method is to add an amplifier stage with gain rolled off at some frequency, represented by Block **G** in Fig. 7.6a. This method has at least three shortcomings:

- If this is to be a low-frequency boost, the magnitude of G has to become 1 at the desired cut-off frequency $f_{cut-off}$ and stay at that value at all higher frequencies, which may not be easy to implement.

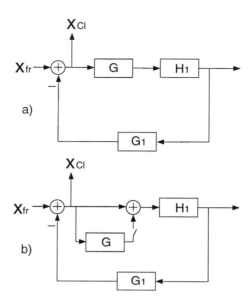

Figure 7.6. Block-diagram illustrating the concept of low-frequency gain boost. a) series boost, b) parallel boost. G and H_1 are transfer functions of electronic amplifiers. The transfer functions of the actuator and the plant are lumped into G_1. The sensor, assumed to have unit transfer function, measures the output of the left-most summing node and feeds into the electronic amplifier(s). X_{fr}: free-running variable (or disturbance), X_{Cl}: residual free-running variable when the loop is closed.

- Adding a stage in series increases the propagation delay experienced by the control signal. If the original loop has high bandwidth, the added delay will most likely decrease the phase margin enough to cause a degradation of the high-frequency performance.

- In real-life situations, there is always some nonzero signal level at the sensor output. If the arrangement of Fig. 7.6a is used to achieve very high gain at low frequencies, the output of the actuator driver is guaranteed to be saturated when the loop is open. For very high gain, the saturation may be so deep that the amplifier will never be able to reach linear operation, which will prevent closing the loop.

All three shortcomings are eliminated by configuring the gain boost in parallel with a unity gain section and adding a switch to the boost path, as in Fig. 7.6b. If the gain of Block G rolls off to a magnitude of one at $f_{cut-off}$ and continues to decrease at higher frequencies, its effect will be felt only for $f < f_{cut-off}$ as desired.

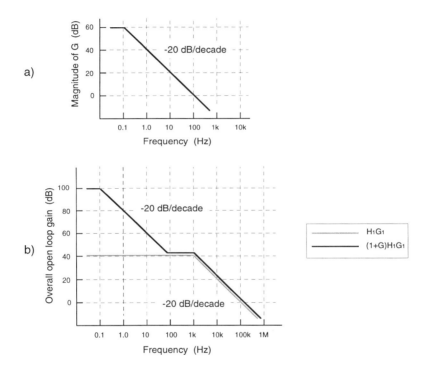

Figure 7.7. Effect of low-frequency gain boost on overall loop gain for systems like the one in Fig. 7.6b. a) Gain magnitude for boost block G. b) Overall open-loop gain. Black: boost switch closed, gray: boost switch open.

The high frequency signals travel straight through via the unit branch and are not delayed by the presence of G. Because of the latter circumstance, this arrangement was called "by-pass" by J. N. Hall, who pioneered the use of this configuration for laser frequency noise suppression. The effect of the by-pass on the overall loop gain is illustrated in Fig. 7.7.

Before lock acquisition, the switch is kept open. In this state, the overall gain is low and deep saturation is less likely, which makes closing the loop possible. Once the loop is closed, the output of the sensor is suppressed by the existing loop gain. In this situation, even with the higher gain provided by the boost, saturation is avoided. The boost can thus be activated by closing the switch. The switch also provides the added convenience of allowing the gain boost to be turned on or off, e. g. for testing, while the system is operating.

1.3. Two Actuators with Nested Loops for Keeping the Optical Path Constant

In order to keep constant the optical path traveled by light in certain high-precision instruments, it is common practice to attach one of the mirrors directing the light to a PZT device like in Fig. 7.8a. The optical path fluctuations are suppressed by applying a suitable signal to the PZT, which in turn moves the mirror and thus keeps the optical path constant. The corresponding closed-loop arrangement is shown in Fig. 7.10a. A difficulty with this setup arises when the PZT, selected to have adequate speed for the application, cannot cover the full range of the disturbance. In the example of Fig. 7.9, the total disturbance extends over a range of ~ 100 μm, peaked at very low frequencies. The Polytek PZT stack P249.10, already mentioned above, is fast enough to address the fastest disturbances in this example, but its range is only 5 μm, far short of the needed 100 μm. The low frequency command signal will attempt to make the PZT move beyond its range, at which point the device will "saturate" and closed-loop operation will break down. Replacing the PZT with an actuator with wider range will usually not be satisfactory, as wide range usually comes at the expense of speed, which in this example would reduce the bandwidth of the closed-loop system and thus force a reduction in gain even at low frequencies; the loss of disturbance rejection may be unacceptable. This is the opposite of the situation described in Section 1.1, where the actuator initially present in the loop had enough range, but not enough speed and had to be supplemented with a faster actuator. It is thus desirable that the PZT be maintained and that the remedy include:

- A means to cover the entire range of the disturbance and
- a reduction of the command signal to the PZT at low frequencies.

Multiple Signal Paths 109

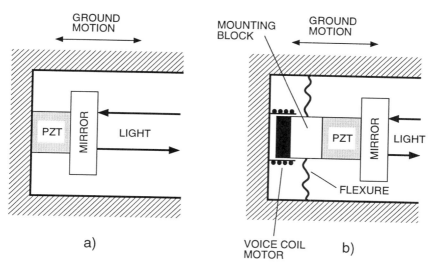

Figure 7.8. Arrangement for suppressing mirror vibration caused by ground motion. a) single actuator, a PZT, b) voice coil added to increase range at low frequencies.

This can be accomplished by mounting the mirror on flexures and adding a voice coil to push the PZT/mirror assembly as in Fig. 7.8b. Voice coils can easily provide correction ranges of several millimeters, which is more than adequate for this example. A control system arrangement which can be used for the voice coil/PZT combination is shown in Fig. 7.10b. The error signal for the voice coil path is the PZT command signal, taken from the output of the PZT compensator G_1. This signal is shaped by $G(s)$, then applied to the voice coil. Going through the algebra, one finds that indeed the PZT command signal is reduced $(1 + G_1 H_1 + G G_1 H_1)/(1 + G_1 H_1) \sim G$ times. If $|G(j\omega)|$ is sufficiently large (larger than 100 μm/5 μm = 20 in this example), the PZT will be desaturated and the loop will work properly. The nested loop configuration of Fig. 7.10b is typical of arrangements where it is necessary to desaturate a fast, narrow-range actuator.

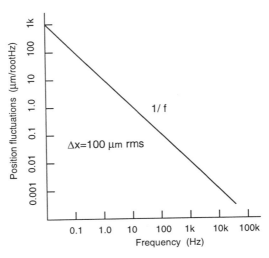

Figure 7.9. Example of displacement fluctuation spectrum affecting the position of the mirror of Fig. 7.8.

2. PARALLEL AND NESTED LOOPS: EQUIVALENCE AND STABILITY

2.1. Equivalence

The three configurations presented as examples In Figs. 7.4, 7.6b, 7.10b can be shown to be equivalent in a topological sense. As shown in Fig. 7.11, the parallel-path diagram of Fig. 7.4 can be transformed into the nested diagram of Fig. 7.10b by using the following steps:

1 Move the G_1 block of Fig. 7.4 in line with the H_1 block. The inner path is unchanged.

2 The outer path now has an added block G_1 in series with the previously existing block G_2. In order to restore the the open-loop response of this path to its initial value, replace G_2 by G_2/G_1.

The resulting diagram, Fig. 7.11b, is formally equivalent to the parallel path diagram of Fig. 7.4. It is also the diagram for a nested loop

Multiple Signal Paths

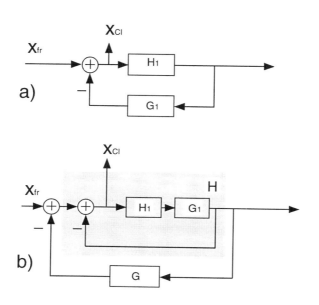

Figure 7.10. Nested loop arrangement for preventing actuator saturation. a) Initial arrangement. b) Initial loop is shown nested inside a second loop using a slow actuator which has sufficient range to ensure system operation.

arrangement with the inner path identical to that of Fig. 7.10b, and with the outer path having response $G = G_2/G_1$.

As shown in Fig. 7.12, the parallel gain-boost arrangement of Fig. 7.6b can be transformed into a nested diagram by going through the following steps:

1. Move the G_1 block of Fig. 7.6b in line with the H_1 block (Fig. 7.12b). The diagram is unchanged.

2. Move the boost block G and its parallel unit-gain path after H_1, G_1 (Fig. 7.12b). The diagram is unchanged.

3. Instead of adding the unit-gain path and G promptly, add G directly to the input path, (Fig. 7.12c).

4. Redraw the diagram slightly, as in Fig. 7.12d.

The diagram of Fig. 7.12d is identical with Fig. 7.10b.

112   *FEEDBACK CONTROL SYSTEMS*

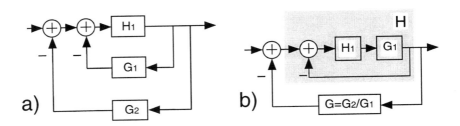

Figure 7.11. Transformation of parallel path diagram into equivalent nested diagram. The notations in these diagrams were chosen in order to make the topological equivalence with the nested loop easier to see. For the sake of further discussion, "1" is assumed to be the faster path.

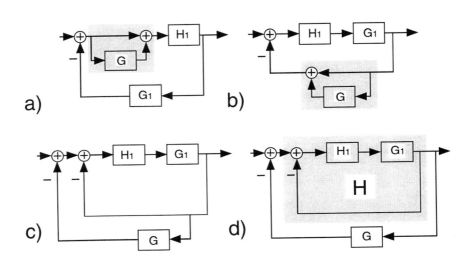

Figure 7.12. Transformation of parallel gain-boost diagram into equivalent nested diagram.

Multiple Signal Paths 113

2.2. Stability Criterion

Nested Configuration
Consider a nested-loop system as in Fig. 7.10b under the following assumptions:

1. H_1, G_1, G are transfer functions of stable systems, i. e. they have no poles in the right-hand side of the s-plane.

2. The inner loop with H_1, G_1 is stable on its own.

3. The unity gain frequency f_1 of G is much lower than the unity gain frequency f_0 of H_1G_1, the open-loop gain of the inner loop, that is $f_1 \ll f_0$. This is reasonable, since the outer loop is added in order to reduce the control effort required of the inner loop at low frequencies, and it does not have to provide any assistance with control at frequencies close to f_0.

Since the inner loop is stable on its own, the only stability concern relates to what happens around f_1, the unity gain frequency of G. For frequencies $f < f_0$, the overall open-loop gain of the system is:

$$L = HG = \frac{H_1G_1}{1+H_1G_1}G \sim G \qquad (7.1)$$

since at these frequencies $H_1G_1 > 1$ by definition. The following stability criterion can thus be used for the nested-loop arrangement:

- The inner path H_1G_1 needs to be stable on its own.
- G shall satisfy e. g. the Nyquist stability criterion.

Parallel Configuration
It was shown above that the parallel path arrangement of Fig. 7.4 is equivalent with a nested loop arrangement where $G \rightarrow G_2/G_1$. The stability criterion for the nested-loop arrangement can thus be rephrased as follows for the case of a parallel-path system:

- The "fast" path H_1G_1 needs to be stable at its high-frequency unity gain frequency f_0, which is also the unity gain frequency for the whole loop.

- G_2/G_1 shall satisfy the Bode stability criterion at the hand-off frequency f_1.[2]

Notes:

1. The above stability criterion requires that the relative phase between G_2 and G_1 be less than 180°, e. g. 150°, at the hand-off frequency f_1. Alternatively, the relative slope should be just less than 12 dB/octave at the hand-off frequency. This is less restrictive than requiring the relative slope not to exceed 6 dB/octave, an ad-hoc criterion sometimes used to design this kind of loop, and therefore is conducive to achieving higher gain at low frequencies.

2. A further implication of the approximate stability criterion for this case is that the fast path does not have to be stable on its own below the hand-off frequency f_1, where overall loop behavior is determined by the slow path.

Parallel Gain Boost

From the equivalence between the parallel gain boost configuration and the nested configuration, stability of the gain boost arrangement requires that the response G of the gain boost block satisfy the Nyquist criterion at the cut-off frequency $f_{cut-off}$, in addition to the unboosted loop with $H_1 G_1$ being stable on its own.

3. THE PID COMPENSATOR

A well known and commonly used arrangement with parallel control paths is the **PID compensator**, with general ideal transfer function:

$$G_{PID} = \frac{\alpha}{s} + \beta + \gamma s \tag{7.2}$$

The first term is an ideal integrator, the second term is a constant coefficient (i.e. a proportional term) and the third term is an ideal derivative.

---

[2]This stability criterion was found by A. Abramovici, J. Harman and M. E. Zucker in 1988 (unpublished), and by B. Lurie, published as "On the zeros of the sum of to impedances in the right-hand plane," *Telecommunications* No. 9, 1962.

Multiple Signal Paths

In practice, the frequency range over which an integrator or a derivative can be implemented is finite, and the transfer function of a real PID compensator is:

$$G_{PID} = \frac{\alpha}{s + 2\pi f_1} + \beta + \gamma \frac{s + 2\pi f_2}{s + 2\pi f_3} \qquad (7.3)$$

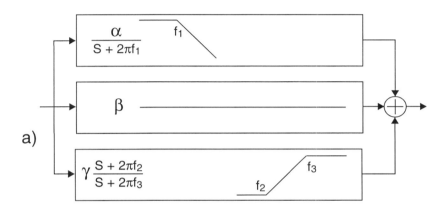

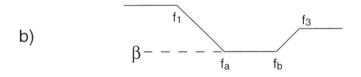

Figure 7.13. PID compensator. a) FCS diagram showing the integrator, proportional, and derivative blocks as parallel control paths. Each block shows the transfer function and the corresponding asymptotic Bode diagram. b) Asymptotic Bode diagram of a PID compensator for a particular choice of the coefficients α, β, γ, resulting in corner frequencies such that $f_1 < f_a < f_2 < f_b < f_3$.

PID compensators (see the diagram of Fig. 7.13a) are attractive to scientists and engineers because of their simplicity and versatility. PID compensator implementation can take the form of a general purpose laboratory instrument which allows the operator to change the coefficients α, β, γ and the corner frequencies f_1, f_2, f_3 by turning front-panel knobs, so that a variety of compensator transfer functions can be realized with a single instrument. An example of a transfer function which can be realized with a PID compensator is shown in Fig. 7.13b.

The slopes characterizing the terms of a PID compensator are either 0 dB/decade or ±20 dB/decade. The corresponding relative phases at the hand-off frequencies between terms are therefore ±90°. According to the hand-off stability criterion developed in the previous Section, the PID compensator hand-offs are thus stable with a robust phase margin of 90°. While this may look like a good thing, it also means that the gain increase towards lower frequencies is not as steep as it could be, if one chose a lower hand-off phase margin, e. g. 30°.

4. CHOOSING A MULTIPLE-PATH CONFIGURATION

While the configurations discussed earlier are formally equivalent from a topological point of view, significant differences arise when it comes to their use in practice. A few practical considerations are listed below.

1. The fact that the ratio between control efforts in a parallel path configuration is approximately $|G_2/G_1|$ (see Fig. 7.11) means that this arrangement can be used to desaturate a fast but narrow-range actuator. This can be of significant advantage in digital control systems, if the fast actuator requires an analog command signal. In this case, strict adherence to the nested configuration would require the analog command signal to be redigitized for the implementation of the desaturation path G.

2. A somewhat subtle point is that one can obtain higher desaturation ratios from parallel path arrangements than from nested loop configurations. This is due to the fact that at low frequencies, where the slow path determines the behavior of a loop with parallel paths, the fast path can have zeros in its transfer function, which for given G_2 can yield a very large $|G_2/G_1|$ ratio. In a nested loop system, on the other hand, the $H_1 G_1$ paths has to be capable of closed-loop operation, albeit over a narrower range of inputs, even in the absence of the desaturation path G, and desaturation is limited to $|G|$. Thus, the nested loop arrangement trades desaturation ratio for more robust closed-loop behavior.

Multiple Signal Paths 117

3 Some parallel path arrangements are hard, if not impossible to implement as nested loops. An example is the laser frequency noise suppression system with transfer functions as in Fig. 7.5. Since the fast path driving the electro-optic crystal has to have zero DC gain, the corresponding command signal is inappropriate for driving the slow actuator (the PZT), which has to correct for slow changes in laser frequency. Thus, the electro-optic crystal branch cannot be implemented as H_1G_1 of Fig. 7.10b. It is also not practical to implement the electro-optic crystal path as G and the PZT path as H_1G_1 in Fig. 7.10b. Indeed, the input signal to G is then derived from a signal which has passed through the entire PZT (slow) part of the loop, and has thus acquired significant propagation delay. The latter will reduce the phase margin of the fast loop to a point where it would seriously decrease its high-frequency stability.

4 The implementation of the low-frequency gain boost of Fig. 7.6b can be purely electronic, as it requires just a gain stage and a summing node. It would be definitely impractical to build it as in Fig. 7.12d, which requires an additional actuator to apply the additional correction on the input disturbance.

Configuration	Main uses	Comments
Parallel paths	Actuators with partial frequency range overlap	Low propagation delay, high desaturation ratio. More design effort needed, less robust in operation than nested loop.
Gain boost	Gain boost over selected frequency range	Low propagation delay
Nested loop	Low-frequency desaturation of fast actuator	Robust operation, intuitive design approach. Lower desaturation ratio than parallel path configuration
PID	General purpose lab compensator. Stabilization of plants with double integrators.	Versatility, robust hand-off stability. Less than optimum performance.

Table 7.1. Comparison between different configurations with multiple control paths.

Given the wide variety of practical considerations influencing the design, it is hard to provide a rigorous criterion for deciding what configuration is optimal for a specific control problem. Instead, one may ask what the main advantages and disadvantages of each configuration are and thus try to see which one is more suitable for a specific application. A crude selection can be made using Table 7.1.

Chapter 8

DIGITAL COMPENSATORS

So far in this book, the discussion assumed that the compensators will be implemented as analog electronic circuits. One radically different approach to compensator design is to digitize the sensor output and implement the compensator as a digital filter. The filter output is then converted to an analog signal for driving the actuator(s).

This Chapter develops a procedure for specifying the components of a digital compensator. Like in the rest of the text, the emphasis is on providing **a way** to obtain the desired result, rather than attempting to achieve optimal performance in a rigorous sense.

Fig. 8.1 overleaf shows a closed-loop system with a digital compensator. Inspection of Fig. 2.1 (p. 11), where the compensator is implemented as a chain of analog amplifiers and of Fig. 8.1 shows that the control system using a digital compensator, which is an example of a **digital control system**, compares to its analog counterpart as follows:

- Both systems include sensors, actuators and actuator drivers. Independent of the topic of this chapter, these components may be purely analog or may include digital technology to a lesser or wider extent.

- The digital compensator displays several new components: the range matching amplifier, the anti-aliasing filter, and the smoothening filter. Although these components are strictly speaking analog devices, they are not needed in conjunction with purely analog compensators, and thus will be considered part of the digital compensator. They are needed for proper interfacing between the digital compensator and the rest of the system (range matching amplifier) and for avoiding

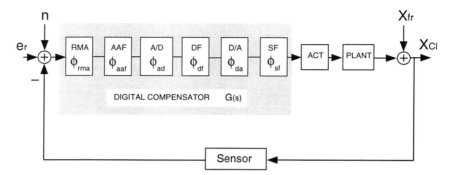

Figure 8.1. Modified version of Fig. 2.1, emphasizing the digital compensator. RMA: range-matching amplifier, AAF: anti-aliasing filter, A/D: analog-to-digital converter, DF: digital filter, which represents the digital implementation of the poles and zeros of the analog compensator transfer function G(s), D/A: digital-to-analog converter, SF: smoothening filter. ϕ_k are the phase shifts introduced by the digital implementation of the compensator at the unity gain frequency f_0 and are added to the phase lag associated with $G(s)$. In particular, ϕ_s, not shown in the diagram, is the phase shift associated with the sampling interval. All input noise up to and including the sensor noise is lumped into one noise term n. The block labeled ACT includes the actuator and the actuator driver.

performance degradation related to discrete time processing (anti-aliasing filter) and to the presence of quantization steps in the digital representation of the data (smoothening filter).

- The digital compensator includes an **analog-to-digital converter**, or A/D. The A/D is a device which samples the analog output of the sensor at a given **sampling rate** and generates a digital representation for each sample. The resolution of the digital representation of the samples is characterized by the number of bits used.

- A component critical to the performance of the digital compensator is the **anti-aliasing filter**. This component ensures that the bandwidth of the analog signal sampled by the A/D is no wider than half the sampling rate $f_s/2$. According to Shanon's sampling theorem, this allows reconstruction of the original signal from the discrete-time samples. If the signal has spectral components at frequencies higher than $f_s/2$, in violation of the sampling theorem, the sampling process

Digital Compensators 121

causes them to be shifted down in frequency by integer multiples of $f_s/2$. The downshifted terms are added to the low frequency components, a phenomenon known as **aliasing**. The samples are thus no longer appropriate for the reconstruction of the original signal.

- The digital filter is the digital equivalent of the analog compensator. It provides the gain and the filtering function needed for implementing the required frequency response. The digital filter is essentially a computer or a digital signal processor executing the code which realizes the overall DC gain and the appropriate frequency response. In either case, hardware and software interfacing with the A/D and D/A are necessary.

- Following the digital filter is a component called **digital-to-analog converter**, or D/A. The D/A is a mixed signal device which converts the output of the digital filter into an analog signal which can be fed to the actuator.

The digital compensator has to perform the same functions as its analog counterpart, namely to provide the DC gain and loop shape that ensure stable operation at a sufficiently high gain and low noise level, as required by the performance specifications. Systems considerations related to analog input noise and loop shape are the same as in the analog case, and will therefore not be repeated here.

Some considerations concerning the use of digital compensators are summarized in Section 1. The quantities relevant in specifying the components of the digital controller are discussed in Section 2. A step-by-step digital controller specification process is outlined in Section 3.

1. WHEN SHOULD ONE USE DIGITAL COMPENSATORS?

There seems to be a widespread belief that the main advantage of a digital compensator versus an analog one consists of the ease in digitally implementing or changing a filter function, as opposed to designing, building and modifying an analog electronic circuit. Most often, when the whole system is considered, this turns out to be an illusion. Indeed, digital compensators come at a cost:

- Fig. 8.1 and the preceding discussion suggest that digital compensators lead to higher system complexity and likely higher cost than purely analog compensators.

- The presence of additional components adds to the propagation delay in the loop, thus decreasing the phase margin and possibly degrading the stability of the system (see Eq. 8.1 on p. 124).

It is a fact, though, that digital compensators are chosen for an increasing number of applications. Here are a few reasons why one should chose a digital compensator:

- Consider a digital compensator when the sensor or the actuator or both are intrinsically digital devices. In this case the system may actually become simpler with the choice of a digital compensator.

- Some systems which are out of reach, e. g. space probes, may require changing the compensator transfer function occasionally. In this case, the compensator transfer function is modified by up-loading the code for the new digital filter.

- When a control system is built into mass-produced articles, improvements in compensator design can be implemented in the entire installed base simply by distributing the new code on floppy disks.

- Analog filters with very high gain are difficult to implement because of the ever present DC offset at the input, which gets amplified with the signal, to a point where it may saturate amplifier output, thus making the system unworkable. Digital filters with very high gain do not have this problem, as their input offset is zero. This advantage may be partially upset by an offset of one least significant bit or more at the first link in the digital chain, the analog-to-digital converter.

- Similarly, digital compensators are intrinsically free of problems associated with drift caused by temperature changes, which may be intolerable for certain sensitive applications.

- Filters with time constants of tens of seconds and longer are easier to implement digitally than to build from analog components. This makes digital compensators particularly attractive for applications where FCSs with low unity gain frequencies are required. The ease of implementing digital filters with long time constants by using slow and thus cheap computers, A/D's and D/As makes this the exclusive province of digital compensators.

- Except for the nonlinearities associated with the quantization step (the least significant bit), round-off errors and the range limit set by finite word length, digital compensators are linear. In contrast, all analog compensators are suffering from some degree of nonlinearity over their entire range, and have the strong nonlinearity resulting from range limitation related to the power supply voltage.

2. ASPECTS OF DIGITAL COMPENSATOR DESIGN

There are many ways to approach the design of a digital compensator. The way described below starts by collecting the following information:

- The analog transfer function $G(s)$ derived in Chapter 6, Section 4, p. 72, Eq. 6.11, which the digital compensator is supposed to realize.

- The range D of the signal at the compensator input.

- The noise spectral density $N(f)$ at the sensor output (Fig. 8.1). One common design requirement is that $N(f)$ be the dominant noise in the system, thus the specification of the digital compensator has to ensure that noise contributions from its components are lower, let's say by a factor 3.

The phase margin associated with the open-loop frequency response $L(j\omega)$ does not take into account the propagation delay of signals through various components peculiar to the digital compensator. In all-analog systems, the propagation delays can usually be made insignificant by choosing adequate fast components and by collocating the sensor with the actuator. In a digital system, the effect of sampling and computation can add a significant phase shift which might affect system stability. Thus, in specifying a digital compensator, one has to allow explicitly for phase margin degradation, by allocating a certain predetermined amount of phase shift to accommodate sampling, computation time, etc. The **phase reserve** ϕ_r is thus defined as the acceptable limit to the degradation of the phase margin ϕ_m, brought about by the use of a digital compensator. For example, if the phase margin designed into $L(j\omega)$ is 50°, one could allocate $\phi_r = 20°$ to allow for delays in the digital compensator. The remaining phase margin of 30° will still ensure robust stability. The design process has to ensure that ϕ_s, the phase lag associated with the sampling process, plus the sum of all propagation-related phase shifts shown in Fig. 8.1 should be bounded by the phase reserve:

$$\phi_r \geq \phi_{rma} + \phi_{aaf} + \phi_s + \phi_{ad} + \phi_{df} + \phi_{da} + \phi_{sf} \qquad (8.1)$$

The material presented in the remainder of Chapter 8 refers to the class of systems where good noise performance is crucial. Thus, it will be assumed, as in Section 6, that $N(f)$ dominates the noise generated by the digital compensator.

2.1. A/D Converter and Anti-Aliasing Filter

Analog-to-Digital Converter Sampling Rate

The finite time interval from sample to sample causes a delay at the output of the digital compensator, which adds to the phase shift built into the required frequency response $G(j\omega)$. The value of this phase shift at frequency f_0, in degrees, is:

$$\phi_s = 180° \cdot \frac{f_0}{f_s} \qquad (8.2)$$

where f_s is the sampling frequency. This relationship comes about as follows:

- The fastest component in the sampled signal is at frequency $f_s/2$, due to the anti-aliasing filter.

- The sampling interval $1/f_s$ corresponds to half of the period of this spectral component, $2/f_s$.

- The phase shift corresponding to a half-period is 180°.

- All other spectral components are at lower frequencies and therefore have longer periods, which in turn corresponds to lower phase shifts over the sampling interval.

According to Eq. 8.2, the minimum sampling frequency required for stable operation is

$$f_s \geq \frac{180°}{\phi_s} \cdot f_0 \qquad (8.3)$$

Digital Compensators

with ϕ_s subject to Eq. 8.1.

Anti-Aliasing Filter

- The anti-aliasing filter has to suppress noise above $f_{cut-off} = f_s/2$ which is otherwise folded onto the lower frequencies of interest. Suppression needs to be sufficient to ensure that noise in the frequency band of interest is not significantly enhanced by aliasing. The exact type of filter depends on the details of the noise spectrum at frequencies above f_s, and has to be determined on a case-by-case basis. In the particular case of white noise, adequate anti-aliasing is provided by a two pole low-pass filter, with the cut-off frequency at $f_s/2$, which introduces a phase shift:

$$\phi_{aaf} = 180° \cdot \frac{f_0}{f_s} \qquad (8.4)$$

at the unity gain frequency of the control system. This phase shift has to satisfy Eq. 8.1. One should be warned, however, that noise is seldom white. Sometimes, noise increases at higher frequencies, requiring a steeper filter cutting in at lower frequencies. If this is the case, Eq. 8.4 is too optimistic, and the phase reserve could be easily depleted just by the presence of the anti-aliasing filter.

- Anti-aliasing filter noise, referred to its input, should not to exceed $N(f)/3$.

Analog-to-Digital Converter Resolution

Signal quantization induces an uncertainty in the value of the signal, which is known only within the least significant bit. This effect can be accounted for by an equivalent additive quantization noise at the A/D input, with variance:

$$\sigma_q = \frac{\delta}{3.46} \qquad (8.5)$$

where δ is the signal range corresponding to one least significant bit. Eq. 8.5 follows commonplace notation practice. σ as used in this Chapter should not be confused with the real part of the Laplace variable, also denoted σ in this book. Quantization noise should not add significantly to the analog input noise σ, e.g. $\sigma_q \leq \sigma/3$. The signal range corresponding to the least significant bit is thus:

$$\delta \approx \sigma \qquad (8.6)$$

The analog noise level σ is calculated as:

$$\sigma^2 = \int_0^{f_s/2} n^2(f) df \qquad (8.7)$$

From Eq. 8.6, the number of bits b needed to properly resolve the input signal is:

$$b_{ad} = \log_2(D/\sigma) = \frac{\log(D/\sigma)}{\log 2} \qquad (8.8)$$

where $D = B \cdot \mathcal{R}_x$ is the range of the analog signal, with B the gain of the sensor and $\mathcal{R}_x$ the range of $x_{fr}(t)$. It is important to specify D conservatively, otherwise a condition of overflow may occur at the analog-to-digital converter, which is likely to result in control system malfunction. In practice, because of noise pick-up at the A/D, one needs to specify a higher resolution than indicated by Eq. 8.8, e. g.:

$$b_{ad} = \frac{\log(D/\sigma)}{\log 2} + 2 \qquad (8.9)$$

For noise with white spectrum, Eq. 8.7 becomes $\sigma^2 = n^2 \cdot f_s/2$. Plugging this into Eq. 8.9 yields the interesting result that, for given noise power spectral density and dynamic range, one can reduce the required number of bits by increasing the sampling rate. If, as is most often the case, noise is colored, one has to actually calculate the integral of Eq. 8.7, and the speed/resolution trade-off is not so obvious. At the most general level, the speed/resolution trade-off is addressed by another Shanon theorem. Some practical aspects of this trade-off are discussed in Appendix C.

Analog-to-digital Converter Propagation Delay
The propagation delay corresponding to the analog-to-digital converter is equal to the time interval between sampling the analog input

Digital Compensators 127

and issuing the digital output. Usually, A/D converters capable of faster sampling rates also have lower propagation delays. This provides additional incentive to chose the fastest analog-to-digital converter which is reasonably priced, once the necessary resolution has been determined. It is worth spending a few more dollars in order to reduce the propagation delay introduced by this part, which will result either in increased robustness or in higher performance for the overall system.

2.2. Range Matching Amplifier (RMA)

Since the sampling rate and resolution specifications for the analog-to-digital converter are likely to be extreme, it is assumed that one will not be able to specify the range independently, and that one will have to live with the range of one among very few available devices. A range matching amplifier is then needed to match the range $D = B\mathcal{R}_x$ of the analog signal to the range D_{ad} of the analog-to digital converter. The basic specifications of the RMA are:

- Gain: D_{ad}/D.
- Output range: D_{ad}, input range: $D = B\mathcal{R}_x$.
- Bandwidth: $BW \gg f_s$, so that the additional phase shift contributed at f_0 is minimized.
- Input referred noise: not to exceed $N(f)/3$.

2.3. Digital Filter Block

Digital Filter Response

Digital filter design can be done in a variety of ways, and the expanse of the phase space available for optimization is daunting. To somewhat narrow down the available options, the following design path is taken here, with substantial loss of generality:

1 As stated earlier, compensator transfer function design is carried out by using common analog design methods, resulting in the analog transfer function $G(s)$.

2. Next, $G(s)$ is expanded in partial fractions:

$$G(s) = \sum_{k=1}^{P} \frac{A_k}{s - s_k} \qquad (8.10)$$

For simplicity, and also because this frequently occurs in practice, Eq. 8.10 is written under the assumption that, besides zeros, the transfer function has only P simple poles, all real and located in the left half of the s-plane. This assumption is made only for the sake of clarity. The presence of complex poles should not substantially affect the conclusions. The residues A_k and the pole positions s_k are then used to write the z-transform of the digital filter in the **impulse invariance** approximation:

$$G(z) = \sum_{k=1}^{P} \frac{A_k}{1 - e^{s_k T} z^{-1}} \qquad (8.11)$$

where $T = 1/f_s$ is the sampling time interval. The impulse invariance method provides a good discrete equivalent of the analog transfer function if the input signal is bandlimited. This condition is satisfied by the presence of the anti-aliasing filter, which ensures that the sampled signal is a faithful representation of the analog input signal.

3. Eq. 8.11 is used as a prescription for implementing what is called a **parallel realization** of the digital filter. This particular realization is chosen here for illustration purposes. Each term in Eq. 8.11 corresponds to a first order finite difference equation

$$y_k[n] = e^{s_k T} y_k[n-1] + A_k x[n] \qquad (8.12)$$

where y_k are samples of the filter output and x_k are samples of the input signal processed by the filter, provided by the A/D. The output of the filter is calculated as:

$$y[n] = \sum_{k=1}^{P} y_k[n] \qquad (8.13)$$

4. Eqs. 8.12,8.13 are used to specify the word length which ensures proper noise performance and the accuracy of the filter, and the computation speed necessary to maintain control system stability.

Word Length

The length of the words used internally by the computer to handle data and filter coefficients determines the accuracy to which the expression in Eq. 8.11, with truncated coefficients, represents the original analog transfer function. Moreover, finite word length causes noise, generated by rounding off the results of additions and multiplications occurring in Eqs. 8.12,8.13. Therefore, the number of bits used to represent numbers needs to be sufficiently high to guarantee that the dominant noise in the system is the one coming with the input signal, and that the filtering matches the transfer function specification to some minimum level of accuracy. In what follows, a floating point representation of numbers is assumed. In the standard IEEE floating point representation, a 32 bit (single format) word uses 2 bits for exponent and mantissa signs, 8 bits for the exponent and 22 bits for the mantissa. A 64 bit (double format) word uses 2 bits for exponent and mantissa signs, 11 bits for the exponent and 51 bits for the mantissa. The word-length related error terms are addressed below.

1. Coefficient truncation

The part of Eq. 8.11 most sensitive to coefficient truncation are the denominators, which describe z-plane poles at

$$z_k = e^{-2\pi f_k/f_s} \approx 1 - 2\pi \frac{f_k}{f_s} \quad (8.14)$$

For s-plane poles much lower than the sampling frequency, the corresponding z-plane pole is close to the unit circle, and thus the behavior of the filter is sensitive to small errors in pole position. In extreme cases, the pole may move outside the unit circle, and thus result in an unstable filter. An ad-hoc criterion used in what follows requires the truncation error to be less than $1/1000$ of the pole frequency, relative to the sampling frequency, i.e. $2^{-b} < 0.001 f_k/f_s$, where b is the length of the binary mantissa. The minimum mantissa length requirement for the equation corresponding to the kth pole then is:

$$b_{m;k} = \frac{3 + \log(f_s/f_k)}{\log 2} \quad (8.15)$$

It is worth noting that the lowest frequency pole, at f_1, sets the most stringent requirement. While Eq. 8.15 will give a rough idea of the precision needed to represent the pole with some arbitrarily chosen level of accuracy, it is a good idea to run a computer test of the digital filter, and check whether it provides a satisfactory representation of the analog counterpart $G(s)$ of the transfer function.

2 Computation noise
Computation introduces noise because of the need to round off numbers in order to fit the word length. A crude estimate of mantissa length which ensures the preservation of signal to noise ratio from input throughout the digital filter, is:

$$b_m = \frac{2\log(D/\sigma) + \log P + \log(f_s/f_1) - 0.98}{2\log 2} \quad (8.16)$$

This formula applies for the P-pole system defined by Eq. 8.11 and takes into account noise generated by evaluating Eq. 8.12. Noise generated by calculating the sum in Eq. 8.13 is negligible. Eq. 8.16 is somewhat pessimistic, as it assumes that all poles have the same effect as the lowest frequency pole at f_1. However, the word length is overestimated only by 1-3 bits, for most practical cases. Eq. 8.16 is only a rough guide, and the noise performance of the digital filter should be tested by running it on the computer with an artificially generated, realistic input signal.

Computation Speed

In order to get a rough idea about how long it takes to evaluate Eq. 8.12, a similar expression was calculated on a SPARC-20 workstation and timed at 0.45 μs. Thus, for a P-pole system, $T_{df} = 0.5 \cdot P$ μs are needed to calculate one point of the sequence $y[n]$. This value is easily scaled to other processors by taking into account the fact that the SPARC-20 is capable of $R \sim 10^8$ operations/second. The corresponding phase shift, in degrees, at the unity gain frequency f_0 of the control system, is:

$$\phi_{df} = 360° \cdot f_0 T_{df} \quad (8.17)$$

Digital Compensators 131

where ϕ_{df} has to satisfy Eq. 8.1.

Output Register Length

The length of the output register should be able to accommodate the full range of the filter output signal, in fixed point representation. Thus, $b_{da} \geq b_{ad}$, with b_{ad} given by Eq. 8.9.

Note

The material of Section 2.3 was included in order to highlight the issues and provide insight into the design of the digital filter block of a digital FCS. As far as actually sitting down and designing one, keep in mind the following:

- Deriving the coefficients of a digital filter which implements a given analog frequency response is a trivial task these days, thanks to easy to use software packages, like the Signal Processing Tool Box for Matlab.

- The ever increasing speed and word length of modern computers and microprocessors obviates most of the frowning about truncation errors, processing speed, etc.

Given the rosy reality of contemporary digital technology, the recommended design process for the digital filter block can be broken up into the following four steps:

1. Select the fastest affordable microprocessor with a minimum word length of 32 bits.

2. Use an existing software tool to derive the coefficients a digital filter realizing the required analog frequency response.

3. Test the frequency response of the filter using a network analyzer, as described on p. 80. Test the noise performance of the digital filter by feeding it an input signal with the expected input noise level and by measuring the output with a spectrum analyzer. These tests require that the digital filter be connected to the A/D and D/A, as in Fig 8.1 on p. 120.

4. Iterate if necessary.

2.4. Digital-to-Analog Converter

The digital-to-analog converter can be specified as follows:

- D/A resolution: commensurate with the output register length.
- D/A rms noise: not to exceed $\nu_b/3$, where ν_b is the voltage step corresponding to one least significant bit at the output register.
- Ringing at frequencies higher than the sampling rate should have amplitude less than ν_b.
- Settling time should less than $T_s = 1/f_s$, the sampling time.
- Bandwidth: same as the sampling frequency.
- Phase shift ϕ_{da} caused by propagation delay subject to Eq. 8.1.

2.5. Smoothening Filter

The smoothening filter is needed in order to eliminate steps and spikes at the D/A output, which result from the changing of the bit contents at the input register of the D/A. The following is an outline of smoothening filter specification.

- The smoothening filter has to suppress noise above $f_{cut-off} = f_s/2$ and high frequency ringing from the digital-to-analog converter, in order to prevent them from propagating through the system and degrading its performance. A two-pole low-pass filter, with the cut-off frequency at $f_s/2$, suppresses unwanted noise by 6 dB at $f_s/2$, and introduces a phase shift:

$$\phi_{sf} = 180° \cdot \frac{f_0}{f_s} \qquad (8.18)$$

at the unity gain frequency of the control system. This phase shift has to satisfy Eq. 8.1.

- The input referred noise of the smoothening filter should not to exceed one third of the digital-to-analog converter output noise.

Digital Compensators 133

3. STEP-BY-STEP SPECIFICATION RECIPE

The material presented in the previous subsection can be summarized in a specification recipe for digital compensators, as outlined below.

1. Collect design requirements, consisting of:
   - Input signal range D
   - Input noise amplitude spectral density $N(f)$
   - Analog counterpart $G(s)$ of digital filter response
   - Tolerances on $G(s)$
   - Unity gain frequency f_0
   - Phase reserve ϕ_r

2. Combining Eqs. 8.2,8.4,8.18, calculate $\phi = \phi_{aaf} + \phi_s + \phi_{sf} = 540 \cdot f_0/f_s$.

3. Calculate the sampling rate f_s by requiring that $\phi = \frac{2}{3}\phi_r$.

4. Calculate the rms input noise σ by using Eq. 8.7.

5. Determine A/D resolution b_{ad}, according to Eq. 8.9.

6. Select specific A/D, with appropriate f_s and b_{ad}.

7. Specify cut-off frequency of two-pole anti-aliasing filter at $f_s/2$.

8. Specify range matching amplifier, according to Section 2.2.

9. Design the digital filter following the four steps outlined in the box on p. 131.

10. Specify the output register length as $b_{da} = b_{ad}$, see Point 5 above.

11. Specify the digital-to-analog converter According to Section 2.4.

12. Specify a two-pole smoothening the filter according to Section 2.5. The cut-off frequency should be $f_s/2$.

13. Calculate the phase shift ϕ_1 due to propagation delay throughout the digital components of the compensator, at f_0.

14. Verify that $\phi+\phi_1 < \phi_r$. If this condition is not met, push the envelope here and there, and make sure it won't blow up somewhere else.

Note: Steps 2,3,13,14 are arbitrarily chosen as one of many ways to satisfy Eq. 8.1 on p. 124.

III

SPECIAL TOPICS

Chapter 9

ACTIVE NULL MEASUREMENTS

In many applications where a parameter needs to be measured, available sensors are adequate in all respects, except that the range of variation of the parameter exceeds sensor range. For example, if a measurement of the angular coordinates of a flying aircraft needs to be carried out, a digital CCD camera is often appropriate, except for its field of view being to narrow to cover all possible positions of the aircraft in the sky. This shortcoming can be sidestepped by using a tracking arrangement like in Fig. 4.1, p. 37 to force the camera to follow the aircraft. As noted in Section 6.3, the correction signal is a measure of the disturbance, which in this case is the motion of the aircraft. Measuring a free-running parameter x_{fr}, consisting of the plant output polluted by a disturbance, by sensing it, using the sensor in a FCS where an actuator nulls the value of x_{fr} and using the correction signal x_c as a representation of x_{fr} is known as **active null measurement**. The system is set in the nulling mode by setting $e_r = 0$. Another situation where an active null measurement may be useful is when the sensor has a strong nonlinearity, which makes calibration difficult. A better measurement can then be carried out if one can arrange for a more linear actuator and if it is possible to include all these parts into the same FCS.

For the purpose of this discussion, Fig. 9.1 shows the diagram of Fig. 2.1 with $E_r = 0$. Following the signals around the loop yields:

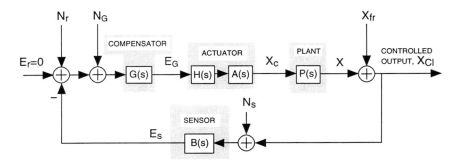

Figure 9.1. Diagram of Fig. 2.1, shown with $E_r = 0$, for the discussion of active null measurements.

$$X_{Cl} = X_{fr}\tfrac{1}{(1+L)} - N_s \tfrac{L}{(1+L)} + (N_r + N_G)\tfrac{L}{B(1+L)}$$
$$PX_c = X_{Cl} - X_{fr} = \left(-X_{fr} + N_s + \tfrac{N_G+N_r}{B}\right)\tfrac{L}{(1+L)} \qquad (9.1)$$

The null measurement consists of first measuring $X_c(f)$ with the loop closed, then calculating $X_{fr}(f) = -P(j\omega)X_c(f)$. In order to perform this operation, the frequency response of the plant $P(j\omega)$ has to be measured with sufficient accuracy. In the context of the active null measurement, the measurement of $P(j\omega)$ is called **calibration**. $-P(f)X_c(f)$ represents $X_{fr}(f)$ with an accuracy (or error) with spectrum $\Delta(f)$. The term $N_s(f)$ is contributed by the sensor and sets a limit to the accuracy of the measurement regardless of whether the sensor is used on its own or in an active null measurement. In other words, it is always true that $\Delta(f) \geq |N_s(f)|$. This and Eq. 9.1 lead to the following bounds for the error Δ of the null measurement:

$$|N_s(f)| \leq \Delta(f) = |X_{fr}(f) + P(j\omega)X_c(f)| \leq$$
$$\leq \tfrac{|X_{fr}(f)|}{|L(j\omega)|} + |N_s(f)| + \left|\tfrac{N_G(f)+N_r(f)}{B(j\omega)}\right| \qquad (9.2)$$

where the approximations $L + 1 \approx L$ and $L/(L+1) \approx 1$ were made. A good quality active null measurement should take full advantage of

Active Null Measurements

the available sensor, i. e. it should be characterized by $\Delta(f) \approx |N_s(f)|$. In other words, the first and third terms in the right hand side of the above equation should not contribute to the error more than $N_s(f)$. This occurs if the following conditions are met:

1. Gain condition: $|L(j\omega)| > \sqrt{2}|X_{fr}(f)|/|N_s(f)|$

2. Noise condition: $|[N_G(f) + N_r(f)]/B(j\omega)| < |N_s(f)|/\sqrt{2}$.

The $\sqrt{2}$ in the two conditions above follows from the fact that, somewhat arbitrarily, the residual x_{fr} and the noise terms were allocated equal contributions to the error Δ of the null measurement.

The box below gives a summary of when and how to use an active null measurement.

Consider an active null measurement when:

- the available sensor is to narrow-range to cover the full variation of $x_{fr}(t)$ **or**

- the available sensor has a strong nonlinearity **and**

- the available actuator has adequate range for canceling $x_{fr}(t)$.

For a good active null measurement, ensure that:

$$|L(j\omega)| > \sqrt{2}|X_{fr}(f)|/|N_s(f)|$$

$$|[N_G(f) + N_r(f)]/B(j\omega)| < |N_s(f)|/\sqrt{2}$$

Chapter 10

TWO SENSORS FOR ONE VARIABLE

In some cases, it may happen that one sensor has adequate low-frequency resolution, but cannot follow fast variations of the variable x, while another sensor is fast enough, but has poor low-frequency resolution. The solution then is to combine the outputs of the two sensors as shown in Fig. 10.1 below. This is another special case of MIMO, called MISO (Multi-Input-Single-Output) which can be handled without having to call upon full-fledged MIMO formalism.

An example is shown in Fig. 10.2.[1] This sensing arrangement can be used as in Fig. 10.1 in order to

- measure the motion of the system with high accuracy over a wide frequency range or

- control the position of the system with high accuracy over a wide frequency range, in the presence of an external disturbance x_d, **and** measure the disturbance in an active null measurement, as described in Chapter 9.

In order to accomplish the above, the main topics which need to be addressed are listed again for convenience:

1 System stability, which will be discussed Section 1.

---

[1] A combination of an accelerometer with an interferometer for optical path monitoring in a stellar interferometer was first used by M. Colavita. The extension of the idea for wideband measurements at low light levels in the interferometer was suggested by G. W. Neat.

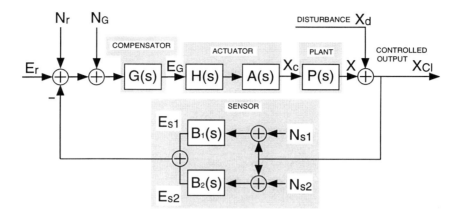

Figure 10.1. Diagram of control system where two sensors provide information on the same variable.

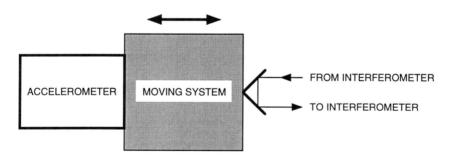

Figure 10.2. Example of two-sensor arrangement for monitoring the motion of a system. The accelerometer attached to the moving system has output proportional to acceleration, $a = \omega^2 x$, which can be converted to motion x by double integration. The system is also coupled to an interferometric sensor via the retroreflector attached to its right hand side. The interferometer yields a signal proportional to motion.

2 Range and bandwidth of actuators: it will be assumed that the actuator in Fig. 10.1 has adequate range and speed.

Two Sensors for One Variable 143

3 Open-loop frequency response. This is relevant for obtaining the required degree of disturbance suppression and reference tracking, or for having the correction signal represent the disturbance with the required degree of fidelity, if the arrangement is used for a null measurement. The convention used on p. 104, that Path 1 is the faster one, and that Path 2 is the slower one, will be adopted here as well. It will be assumed that the open loop gain for the fast path, $L_1 = B_1 GHAP$ is sufficient to ensure proper disturbance suppression/reference tracking at "high" frequencies, according to Eq. 6.10, p. 68.

4 Noise, which will be discussed in Section 2.

1. STABILITY OF CONTROL SYSTEMS WITH TWO SENSORS

The arrangement of Fig. 10.1 is a typical parallel path configuration, as described in Chapter 7. The corresponding stability criterion, p. 113, asks that

$$\frac{L_2}{L_1} = \frac{B_2}{B_1} \quad \text{obey the Nyquist criterion.} \tag{10.1}$$

In the example of Fig. 10.2, the interferometer output is proportional with the motion x, therefore its transfer function is a constant, $B_2(s) = \alpha$. The accelerometer output is proportional to acceleration, therefore in terms of motion its output is $\beta \omega^2$, with β a constant, and the corresponding transfer function is $B_1 = \beta s^2$. The choice of subscripts reflects the fact that the accelerometer is increasingly inefficient towards lower frequencies because of the ω^2 factor, which leads to signal-to-noise less than one below a certain frequency. Thus, the accelerometer is the part of the loop which works better at "higher" frequencies. With these transfer functions, $B_2/B_1 \propto s^{-2}$. This expression has a phase angle of -180°, which causes hand-off instability. The remedy is to supplement the accelerometer with an integrator, which has transfer function s^{-1}. Then, $B_1 = \beta s$, which leads to $B_2/B_1 \propto s^{-1}$. This frequency response function has a phase angle of -90° at all frequencies, which ensures hand-off stability, according to the criterion on p. 113.

144   FEEDBACK CONTROL SYSTEMS

2. NOISE CONSIDERATIONS

Solving for $X_{Cl}(f)$ and $X_c(f)$ following the diagram of Fig. 10.1 yields:

$$X_{Cl}(f) = -\tfrac{\lambda}{1+\lambda}N_{s1}(f) - \tfrac{1}{1+\lambda}N_{s2}(f)$$
$$P(j\omega)X_c(f) = -[X(f) + X_d(f)] + \tfrac{\lambda}{1+\lambda}N_{s1}(f) + \tfrac{1}{1+\lambda}N_{s2}(f)$$
(10.2)

where $L \gg 1$ and $\lambda(f) = L_1(j\omega)/L_2(j\omega) = B_1(j\omega)/B_2(j\omega)$. In order to obtain the simple expressions above, E_r, N_r and N_G were set to zero. This does not change the conclusions concerning sensor noise in this arrangement, since it is assumed, as usual, that sensor noise dominates over all other noise contributions in the system.

In order to make the discussion concrete, it will be assumed that:

- Interferometer noise has flat spectrum spectrum with amplitude equal to 1 nm/$\sqrt{\text{Hz}}$.

- The accelerometer output noise has flat spectrum, which means that the input referred noise behaves like ϵ/ω^2. In this example, ϵ will be chosen such that accelerometer noise equals interferometer noise at 10 Hz, as shown in Fig. 10.3.

With these assumptions, the amplitude of the noise term in Eq. 10.2 has been calculated for three values of the frequency where control is handed off from the interferometer to the accelerometer: 1 Hz, 10 Hz, and 100 Hz. The results, shown in Fig. 10.3, illustrate the fact that in the two-sensor configuration the noise affecting system output is never larger than the highest of the sensor noise contributions. In other words, at any given frequency, the output noise is less than or equal to the noise contributed by the noisier of the two sensors. Thus, the use of two sensors makes it possible to reduce the output noise at some frequencies, at the expense of increasing it at other frequencies. The resulting noise spectrum can be made to be closer to flat, which is desirable in some applications.

Two Sensors for One Variable 145

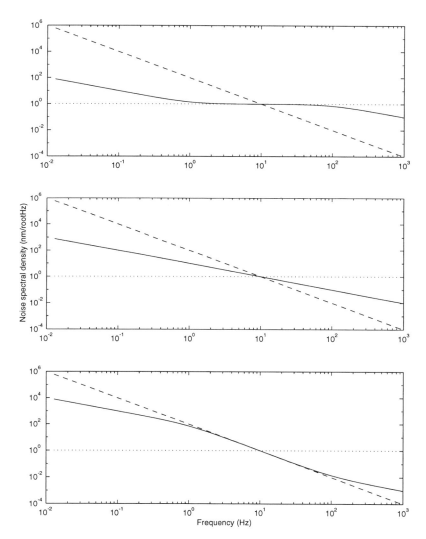

Figure 10.3. Equivalent sensor noise for two-sensor arrangement. Dotted: interferometer noise, dashed: accelerometer noise, solid: equivalent sensor noise affecting $X_{Cl}(f)$. The effect was calculated for three accelerometer/interferometer hand-off frequencies. Top: 100 Hz, middle: 10 Hz, bottom: 1 Hz. In this example, the noise contributions from the two sensors are equal at 10 Hz.

Chapter 11

FLEXIBLE ELEMENTS AND STABILITY

Practical control applications sometimes require that an object be moved. In the example of Fig. 11.1a, a mirror needs to be moved in order to maintain constant the optical path between it and another mirror. Changes in the optical path can be measured with an interferometric gauge. A FCS can be arranged by using the interferometer output as the error signal, which is processed by a compensator and then used to drive a piezo-electric (PZT) actuator supporting one of the mirrors. What can make this kind of arrangement tricky is the fact that the structure which connects the two mirrors is not infinitely rigid. When the PZT moves the mirror, a reaction force acts on the structure. If control is effected at frequencies at or close to structural resonances, the amplitude and phase of the open-loop frequency response is changed and instability may result. The negative effect of structure flexibility on FCS stability and some ways to mitigate it are discussed below.

1. EFFECT OF STRUCTURE FLEXIBILITY ON LOOP RESPONSE

Any structure usually has many resonances, their number increasing rapidly with increasing frequency. Typical table-top optical mounts have their lowest resonances at ~ 100 Hz. In order to understand how resonances impact stability, the effect of a single resonance described by a mass connected to a spring-dashpot combination will be analyzed (see

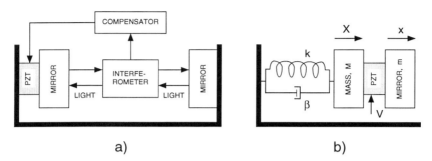

Figure 11.1. Arrangement for holding the optical path between two mirrors constant. a) diagram of closed-loop system, consisting of mirrors, interferometric distance gauge and piezo-electric actuator (PZT). b) representation of structural flexibility by a mass M attached to a spring/dashpot combination. k is the spring constant and β is the damping coefficient.

Fig. 11.1b). The spring is characterized by the spring constant k, and the dashpot represents the damping intrinsic to the structure, described by the damping coefficient β. The PZT, which will be assumed rigid and massless, changes its length as $x_{\text{PZT}} = x - X = k_{PZT}V$, when a voltage V is applied between its two faces, where k_{PZT} is the piezo-electric constant, measured in m/V. Using Newton's second Law, the equations describing the motion of this arrangement are:

$$x = X + k_{PZT}V$$

$$M\frac{d^2 X}{dt^2} = -m\frac{d^2 x}{dt^2} - 2\beta\frac{dX}{dt} - kX \quad (11.1)$$

The variable of interest is x. Solving for x and substituting $d/dt \to s$, where $s = \sigma + j\omega$ is the Laplace variable, one obtains the transfer function from change in PZT length to mirror motion:

$$\frac{x(s)}{x_{\text{PZT}}(s)} = \frac{1}{1 + \frac{m}{M}\frac{s^2}{s^2 + 2\gamma\omega_0 s + \omega_0^2}} \quad (11.2)$$

Flexible Elements and Stability 149

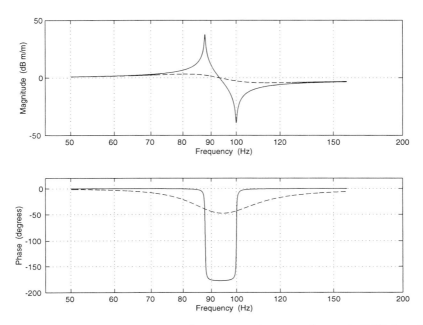

Figure 11.2. Frequency response of mirror motion x to change in PZT length, in units of m/m. This example assumes that in the absence of the mirror, the resonance of the structure is at 100 Hz, and that $m/M = 0.3$. **Solid line:** low damping, $\gamma = 0.0016$. **Dotted line:** high damping, $\gamma = 0.16$.

where the resonance of the structure, without the mirror attached to it, is at frequency $\omega_0/2\pi$, and $\gamma \stackrel{\text{def}}{=} \beta/M\omega_0$.

A Bode plot calculated for $m/M = 0.3$ is shown in Fig. 11.2. While this plot gives one a good idea of the effect of a resonance on the PZT/mirror response, the picture is oversimplified in that real structures have multiple resonances. The example of a fairly damped resonance, with $\gamma = 0.16$ has been added to Fig. 11.2 for completeness. In practice, unless explicit steps to provide damping are taken, metallic structures tend to be lightly damped, with γ typically in the range 0.001-0.01.

2. RESONANCE-INDUCED INSTABILITY AND WAYS TO PREVENT IT

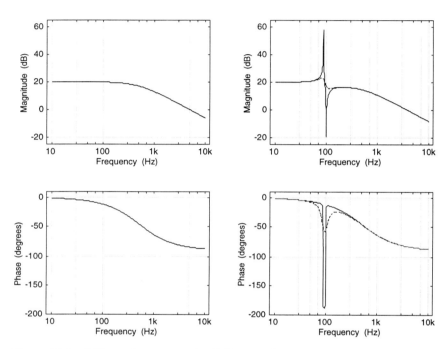

Figure 11.3. Effect of structure flexibility on the pathlength stabilization arrangement of Fig. 11.1. **Left:** Example of open-loop gain with a compensator having a single pole at 500 Hz, assuming flat interferometer response and an ideally rigid structure. Unity gain frequency is 5 kHz and the phase margin is 90°, indicating that this is a stable system. **Right:** open-loop response for the same compensator, including the effect of structural flexibility shown in Fig. 11.2. Due to the sharp spike cause by the lightly damped resonance (solid line) resonance, there is a cross-over at 100 Hz, where there also is an 8° phase deficit, which would result in oscillation. For high damping (dotted line), the wiggle caused by the resonance is too shallow to cause a cross-over. Stability of the system is thus not impaired.

The effect of a structural resonance on FCS stability is illustrated in Fig 11.3. The assumption has been made that a very simple open-loop gain has been implemented with the arrangement of Fig. 11.1. When the effect of the resonance is taken into account, the open-loop gain takes on the behavior shown by the right hand side Bode plot in Fig. 11.3. The lightly damped resonance causes a cross-over at 100 Hz, where it also add a substantial phase shift, leading to an overall phase deficit of 8°. The Nyquist stability criterion is thus violated, and the system will oscillate.

The strongly damped resonance, on the other hand, looks like a small blip on the unperturbed open-loop gain. No additional cross-over at resonance occurs in this case, and the system is stable. Damping is thus a way to prevent structure flexibility from causing FCS instability.

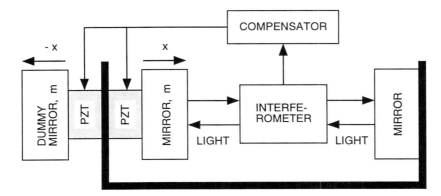

Figure 11.4. Decoupling of PZT/mirror system from structure resonances by **momentum compensation**, sometimes called **reactuation**. A second PZT and a dummy mirror of equal mass are added at the left of the arrangement, back-to-back with the PZT and the mirror. The compensator output is applied to both PZTs, so that the mirror and the dummy are moving in opposite directions by the same amount. The net reaction force applied to the structure can thus be reduced considerably, with a corresponding decrease in the amplitude of the resonance-induced spike in the PZT length change to mirror displacement frequency response.

Another way to reduce the effect of a structural resonance on the open-loop gain is shown in Fig. 11.4. The idea is to duplicate the mirror/PZT assembly and mount the two identical copies back-to-back. At the same time, the compensator output is applied to both PZTs. Then, when the original mirror moves by x, the dummy mirror moves by the same amount in the opposite direction. The resulting forces applied to the structure are equal in magnitude and of opposite sign and therefore cancel each other. As a result, the structural resonance is not excited. In practice, cancellation is never perfect, however a factor 30-100, i.e. 30-40 dB, is easily achieved. This level of resonance attenuation would prevent occurrence of a cross-over at 100 Hz in Fig. 11.3 and thus restore system stability. Like damping, this method does not rely on accurate knowledge of resonance parameters.

Yet another way of reducing the effect of the resonance is to supplement the compensator with a filter which effectively compensates the resonance. While elegant in principle, this method requires accurate knowledge of resonance parameters. Moreover, if the frequency of the resonance changes with time, e. g. as a result of materials aging, or as a function of temperature, the filter needs to be retuned accordingly, which may be impractical. This method is not recommended.

Prevent instability due to structural resonances by:

- Adding damping to the structure.

- Decoupling the actuators from the structure by momentum compensation (reactuation).

Appendix A
Poles and Zeros

Poles

Consider the RC network of Fig. A.1. Assume the input is a sine-wave $e_i = \sin \omega t$.

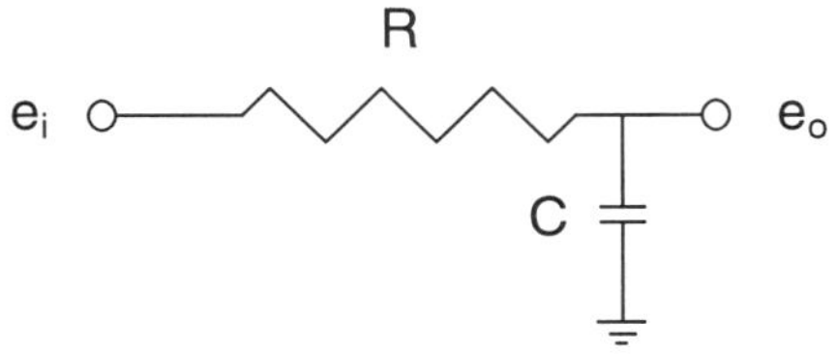

Figure A.1. RC network used to illustrate the concept of pole.

The differential equation which describes this system is:

$$RC\frac{de_o(t)}{dt^n} + e_o(t) = \sin \omega t \tag{A.1}$$

with the solution:

$$e_o(t) = \frac{s_0}{\sqrt{s_0^2 + \omega^2}} \sin(\omega t - \phi) \tag{A.2}$$

where $\phi = \tan^{-1}(\omega/s_0)$, and $s_0 = 1/RC$. The output is a phase-shifted version of the input, with $\phi = -45°$ for $\omega = -s_0$ and $\phi \to -90°$ for $\omega \gg s_0$. For $\omega > s_0$, it is also attenuated at the rate of a factor 10 per decade of frequency (20 dB/decade). In electrical engineering terms, the network of Fig. A.1 is a low-pass filter. The Laplace transform of $e_o(t)$ is obtained from Eq. A.1 by replacing $d/dt \to s$, $\sin \omega t \to \mathcal{L}(\sin \omega t)$ ($\mathcal{L}()$ stands for "Laplace transform of") and solving the algebraic equation. The result is:

$$E_o(s) = \frac{s_0}{s_0 + s} \mathcal{L}(\sin \omega t) \tag{A.3}$$

The transfer function $s_0/(s_0+s)$, which describes the network, has a real **pole** at $-s_0$ in the complex s-plane. For physical frequencies, i.e. for $s = i\omega$, one finds that the amplitude of the input sine-wave is attenuated when $\omega > s_0$, as in Eq. A.2. The imaginary unit multiplying ω describes a 90° phase lag at frequencies well above s_0. The Bode plot for this system is shown in Fig. A.2

Another case of interest is the driven harmonic oscillator, described by the second order equation:

$$\frac{d^2x(t)}{dt^2} + 2\gamma \frac{dx(t)}{dt} + \omega_0^2 x(t) = \sin \omega t \tag{A.4}$$

with the solution:

$$x(t) = Ae^{(-\gamma+i\omega_1)t} + Be^{(-\gamma-i\omega_1)t} + \frac{\sin(\omega t + \phi)}{\sqrt{(\omega^2 - \omega_0^2)^2 + 4\gamma^2\omega^2}} \tag{A.5}$$

where $\phi = \tan^{-1}\left[2\gamma\omega/(\omega^2 - \omega_0^2)\right]$. $\omega_1 = \sqrt{\omega_0^2 - \gamma^2}$ is real for small γ. The first two terms in Eq. A.5 vanish after sufficient time has passed. The remaining term is attenuated for $\omega > \omega_0$, at a rate of ω^{-2}, i.e. a factor 100/decade (40 dB/decade). For $\omega = \omega_0$, $\phi = -90°$ and, for $\omega \gg \omega_0$, $\phi \to -180°$. The Laplace transform of $x(t)$ is:

$$X(s) = \frac{S(s)}{(s - s_1)(s - s_2)} \tag{A.6}$$

The transfer function $(s-s_1)^{-1}(s-s_2)^{-1}$, which describes the harmonic oscillator and has two complex conjugate poles at $s_{1,2} = -\gamma \pm i\sqrt{\omega_0^2 - \gamma^2}$,

Appendix A: Poles and Zeros 155

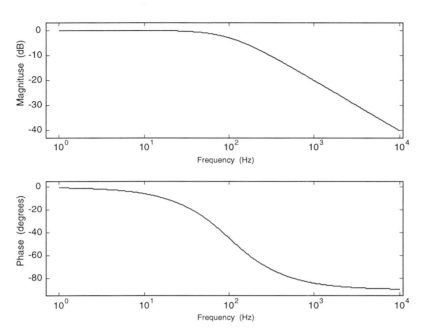

Figure A.2. Bode plot for the RC network of Fig. A.1.

displays the ω^{-2} behavior above ω_0 and the phase lag already seen in the solution of the differential equation.

Differential equations with order higher than 2 are more difficult to integrate than the examples considered above. However, the denominator of the corresponding Laplace transforms is a polynomial consisting exclusively of factors like the denominators in Eqs. A.3,A.6, as long as the systems are described by linear differential equations with constant coefficients. Thus, the corresponding Laplace transforms will have single real poles or pairs of complex conjugate poles.

The preceding discussion of s-plane poles associated with systems described by linear differential equations with constant coefficients is consistent with the following statements, valid for either type of poles:

1 The presence of a pole is associated with low-pass filter behavior.

2 For real poles, the cut-off frequency of the system, i.e. the frequency at which the low-pass filtering behavior sets in, is equal to the position of the pole.

3 For a complex pole, the cut-off frequency is equal to the imaginary part of the pole.

4 For either type of pole, the roll-off, i.e. the rate of attenuation above the cut-off frequency, is 20 dB/decade/pole.

5 Poles introduce a phase lag:

- $\phi = -45°$/pole at the position of the pole.
- $\phi = -90°$/pole at frequencies far above the position of the pole.

Zeros

The presence of poles in the transfer function associated with Eqs. 3.1, A.1, A.4 is related to the derivatives of the unknown function. If derivatives of the forcing function in the right-hand side of the differential equations are present, the transfer function will have a polynomial in the numerator. The roots of the latter are called zeros of the transfer function. Similar to the case of poles, the following statements can be made regarding zeros:

1 The presence of a zero is associated with high-pass filter behavior.

2 For real zeros, the set-in frequency of the system, i.e. the frequency at which the high-pass filtering behavior sets in, is equal to the position of the zero.

3 For a complex zero, the set-in frequency is equal to the imaginary part of the zero.

4 For either type of zero, the roll-up, i.e. the rate of amplification above the set-in frequency, is 20 dB/decade/zero.

5 Zeros introduce a phase lead:

- $\phi = 45°$/zero at the position of the zero.
- $\phi = 90°$/zero at frequencies far above the position of the zero.

Both the magnitude and the phase shift of the frequency response are readily measured quantities. The properties of poles and zeros, listed above, can thus be used to gain insight about the number of poles and zeros of the system at hand, by inspection of the magnitude/phase plots.

Finally, it can be loosely stated that the steeper the roll-off, the larger the related phase lag and, on the other hand, the steeper the roll-up, the larger the related phase lead. At a deeper level, this relates to causality, via dispersion relations. This often causes difficulties in the design of FCSs, when the desire for fast gain roll-off around and above unity gain frequency is frustrated by instability caused by the associated high phase lag. Steep filtering above unity gain is desirable in order to provide passive suppression of electronic noise in a frequency range where the loop is effectively open and thus no noise attenuation by the FCS is available.

Appendix B
Stability of Operational Amplifiers

Since the advent of the integrated circuit (IC), amplifier design has become accessible to a large number of users who are interested in amplifying their signals and less in the nitty-gritty of electronics design. This is possible due to the development of IC opamps, a shorthand for operational amplifiers. Opamps are basically differential amplifiers with very large gain, typically 10^5 or even higher. By using an opamp in a feedback arrangement, as shown in Fig. B.1, one can design an amplifier with specified gain by using a few simple rules, without the need to know any of the detailed workings of the opamp. This is possible because, as long as the gain of the opamp can be considered infinitely high, the corresponding feedback amplifier has, to a good approximation, the following ideal characteristics:

1 The inputs draw no current. In other words, the input impedance is infinite at both inputs, marked with "+" and "-" in Fig. B.1.

2 $v_+ = v_-$.

3 The output impedance is zero.

Yet with all the apparent simplicity, it often happens that amplifier circuits like the one in Fig. B.1 tend to oscillate; in some cases the data sheet of the opamp indicates that stability requires that the gain of the feedback amplifier has to be higher than 5 or 10, which precludes its use in some applications, for example as a unity gain buffer. The object of this Appendix is to analyze the effect of the closed loop on the output characteristics and on the stability of feedback amplifiers which are based on operational amplifiers. The validity of the above ideal characteristics as a function of the open-loop gain A and of the amplifier design will become apparent in the process.

This Appendix covers only system stability aspects of feedback amplifiers using opamps. A comprehensive treatment of operational amplifiers can be found in the classic book by Horowitz and Hill.[1]

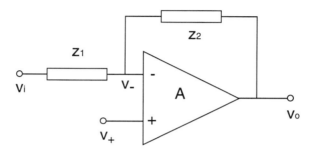

Figure B.1. Inverting amplifier using an idealized operational amplifier. A is the very high gain of the opamp.

1. OPAMP AS FEEDBACK AMPLIFIER

The schematic in Fig. B.2a shows a more detailed representation of a feedback amplifier built around an opamp. The output resistance of the opamp, typically $\sim 100 - 1000\ \Omega$, is explicitly shown. Other real world aspects of amplifier design, which do not bear directly on stability, will still be disregarded:

- Different gains for the inverting and noninverting inputs.

- Finite voltage range at the inputs and at the output.

- Various offsets which add to the input signal and cause an error at the output.

- Noise.

- Temperature dependence of various parameters.

---

[1] Paul Horowitz, Winfield Hill, The Art of Electronics, Cambridge University Press, 1989.

Appendix B: Stability of Operational Amplifiers 161

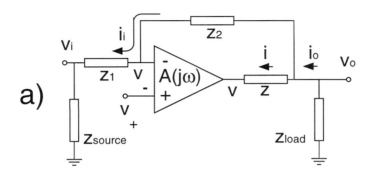

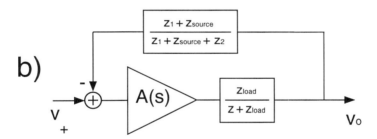

$$L(s) = A(s) \cdot \frac{Z_1 + Z_{source}}{Z_1 + Z_{source} + Z_2} \cdot \frac{Z_{load}}{Z + Z_{load}} \approx A(s) \cdot \frac{Z_1}{Z_1 + Z_2} = \frac{A}{A_{noninv}}$$

Figure B.2. a) a more realistic representation of an amplifier using an opamp. This arrangement differs from the one of Fig. B.1 in that the opamp gain A has a frequency dependence and is **finite**. Also, the opamp output impedance z is explicitly shown. b) FCS representation of the opamp-based feedback amplifier, which will be used for stability analysis. The signal v_i at the inverting input has been set to zero, and the signal v_+ is shown as a command/reference input. While this choice has been made in order to emphasize the FCS nature of the circuit, neither signal matters, as far as stability is concerned. The approximation for $L(s)$ holds for the usual situation where the source impedance is very small and the load impedance is very large.

In order to derive the gain of the amplifier in Fig. B.2a, it will be assumed that $z_{source} = 0$, $z_{load} = \infty$, in other words the amplifier is driven by a source with zero output impedance and it drives a load with infinite impedance. Then, $i = -i_i$ and:

$$v = A(v_+ - v_-)$$
$$\frac{v_- - v_i}{z_1} = \frac{v_o - v_-}{z_2} = \frac{v - v_o}{z} \tag{B.1}$$

Solving these equations yields:

$$v_- = v_+ - \frac{v}{A}$$

$$A_{\text{noninv}} = \frac{v_o}{v_+} = \frac{z_1 + z_2}{\frac{z_1 + z_2 + z}{A} + z_1} \quad ; \quad A_{inv} = \frac{v_o}{v_i} = \frac{\frac{z}{A} - z_2}{\frac{z_1 + z_2 + z}{A} + z_1} \tag{B.2}$$

The output impedance for the amplifier of Fig. B.2 is derived by using the common definition $z_o = v_o/i_o$ with the **inputs shorted**, i.e. $v_i = v_+ = 0$. From Fig. B.2, $i_o = i + i_1$, $v_- = v_o z_1/(z_1 + z_2)$ and $v = -Av_-$. Therefore:

$$z_o = \frac{z}{A} \cdot \frac{z_1 + z_2}{\frac{z_1 + z_2 + z}{A} + z_1} \approx \frac{z}{A} \cdot \frac{z_1 + z_2}{z_1} = \frac{z}{L} \tag{B.3}$$

where the approximation holds for $|z_1| \gg |(z + z_1 + z_2)/A|$ and the open-loop gain L is defined in Fig. B.2. Note that the higher the feedback amplifier gain A_{noninv}, the lower the open-loop gain L. For $A \to \infty$, Eqs. B.2, B.3 become:

$$v_+ = v_-$$

$$A_{\text{noninv}} = \frac{z_1 + z_2}{z_1} \quad ; \quad A_{\text{inv}} = -\frac{z_2}{z_1} \tag{B.4}$$

$$z_o = 0$$

Appendix B: Stability of Operational Amplifiers

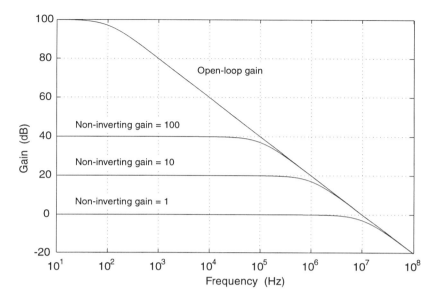

Figure B.3. Relationship between amplifier gain and bandwidth. Because a real-life opamp has finite gain which rolls off with frequency, the bandwidth of the feedback amplifier decreases when its gain is increased.

Note on Feedback Amplifier Bandwidth
When designing amplifiers, one should always keep in mind that Eqs. B.4 hold for $A \to \infty$. In practice, the open-loop gain of the opamp is very high at lower frequencies, then rolls off, typically at a rate of 20 dB/decade as shown in the example of Fig. B.3. When the finite character of the gain $A(j\omega)$ and its frequency dependence are taken into account using Eq. B.2, one obtains the non-inverting gain curves shown in Fig. B.3. The message of this example can be summarized as follows:

- The gain of the non-inverting feedback amplifier cannot be higher than the open-loop gain of the opamp.

- The bandwidth of the non-inverting amplifier is equal to the frequency where its gain becomes equal to the open-loop gain of the opamp.

- Choosing a higher gain for the noninverting amplifier leads to a narrower bandwidth.

- The gain of the non-inverting amplifier does not depend on the details of the open-loop gain of the opamp, as long as $A_{noninv} \ll |A(j\omega)|$.

From Eq. B.2 it follows that similar considerations apply to the inverting amplifier.

2. UNITY GAIN STABILITY AND COMPENSATION

As shown in Fig. B.2, the feedback amplifier built using an opamp can be represented as a FCS, with open-loop transfer function:

$$L(s) = \frac{A(s)}{A_{\text{noninv}}(s)} \qquad (B.5)$$

Thus, $A_{\text{noninv}} = 1 \to L = A$, in other words, when the feedback amplifier has unity noninverting gain, the open-loop gain of the FCS is equal to the opamp gain.

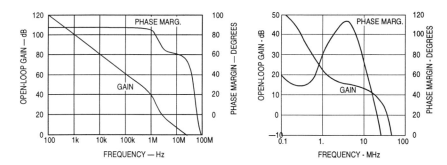

Figure B.4. **Left:** Example of Bode diagram for an opamp which is unity gain stable. The phase margin is $\approx 60°$, indicating that if L is as shown, that is the feedback amplifier is unity gain, the arrangement will be stable. **Right:** Bode diagram showing phase deficit i.e. phase shift in excess of $180°$ at the unity gain frequency. This opamp can not be used for a unity gain amplifier. However, choosing $A_{\text{noninv}} = (z_2 + z_1)/z_1 = 12$ dB would lower the gain curve by 12 dB, which would move the unity gain frequency to a lower value and result in a phase margin of about $45°$, thus ensuring amplifier stability.

Appendix B: Stability of Operational Amplifiers

As for any other real system, above a certain frequency the gain of an opamp has to decrease and the phase shift has to increase with frequency due to stray capacities and parasitic inductances. For a feedback amplifier with unity gain, stability rests on the Nyquist criterion being satisfied by the opamp gain A. Fig. B.4 shows two examples of opamp frequency response. For the Bode plot on the left, the phase margin is about $60°$, satisfying the Nyquist stability criterion. The corresponding feedback amplifier can therefore be stable at unity gain; it is said that the amplifier is **unity gain stable**. For the Bode plot on the right, the phase lag at the unity gain frequency is substantially higher than $180°$, thus the corresponding feedback amplifier cannot be unity gain stable. Stability can be achieved by selecting $A_{\text{noninv}} = (z_2 + z_1)/z_1 > 1$, which causes the open-loop gain L to decrease. This leads to a lower unity gain frequency, with the possibility of improved phase margin. In the example under consideration, choosing $A_{\text{noninv}} = 4$ would move the unity gain frequency to ≈ 10 MHz, which would make the phase margin $\approx 45°$, ensuring feedback amplifier stability.

Some opamps which are not unity gain stable provide the user with the option to **compensate** the circuit for unity gain stability. This option is implemented by providing terminals connected to intermediate points in the opamp internal circuitry. The manufacturer provides a table with values of capacitors which ensure stability for a range of feedback amplifier gains, often including unity. Usually this lowers the opamp gain and possibly also provides a phase lead at the new unity gain frequency.

3. DRIVING CAPACITIVE LOADS

One often finds that an otherwise stable feedback amplifier bursts into oscillations when one attempts to use it to drive capacitive loads. Fig. B.5 shows the corresponding feedback amplifier model. Fig. B.6 illustrates the effect of a capacitive load on the open-loop gain of the feedback amplifier. In this example, a moderate capacitive load, 5 nF, leads to amplifier instability, by adding a pole (see appendix A) to the open-loop gain of the feedback amplifier. The phase lag associated with the pole erodes the phase margin of the closed-loop system and causes the amplifier to oscillate. Oscillation can be prevented by inserting a

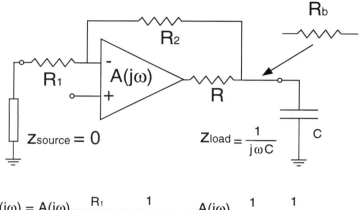

$$L(j\omega) = A(j\omega) \cdot \frac{R_1}{R_1 + R_2} \cdot \frac{1}{1 + j\omega CR} = A(j\omega) \cdot \frac{1}{A_{noninv}} \cdot \frac{1}{1 + jf/f_c}$$

Figure B.5. Model of feedback amplifier driving a capacitive load. The **open-loop output resistance** of the opamp forms a low-pass filter with cut-off frequency $f_c = 1/(2\pi RC)$ with the capacitive load. Thus 45° of phase shift at f_c and 90° at higher frequencies are added to $L(j\omega)$, causing instability. Even though the effective output resistance seen by the load is $R \cdot A_{noninv}/A \ll R$, it is R that determines the location of the destabilizing pole. The fix is to insert a buffer resistor R_b between the output of the amplifier and the capacitive load, as shown.

buffer resistor between the output of the amplifier and the capacitive load, as shown in Fig. B.5.

This works because adding the buffer resistor R_b adds high-pass filter behavior to the open-loop gain. As shown in Appendix A, the zero in this filter introduces a phase lead, which compensates the lag introduced by the low-pass filter. The addition of R_b transforms the low-pass filter consisting of the opamp output resistance R and the load capacitor C into a lag-lead filter (see Fig. 6.10 on p. 88). Some phase margin is restored and, with it, amplifier stability. It can be said that R_b isolates the amplifier output from the capacitive load. Two examples of 5 nF loads are:

- A piezoelectric disk ∼ 10 mm in diameter and 2 mm thick made of PZT, a common ferro-electric material.

- 50 m of RG-58 coaxial cable, frequently encountered in Ethernet computer networks.

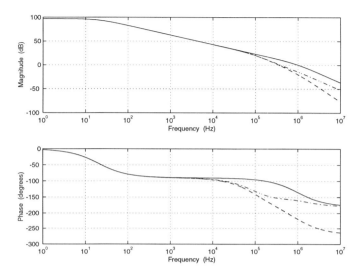

Figure B.6. Effect of capacitive output loading on the open-loop gain of the feedback amplifier, for an opamp with $R = 300\ \Omega$. **Solid:** unloaded amplifier, with $L = 1$ at 1 MHz and 45° phase margin. **Long dashes:** 5 nF load on output, leading to $L = 1$ at 360 kHz and 4° phase deficit, which causes amplifier instability. **Short dashes-dots:** capacitive load compensated with $R_b = 50\ \Omega$ which causes $L = 1$ at 390 kHz and 26° phase margin, thus restoring amplifier stability.

4. INPUT CAPACITANCE AND PHOTODIODE PREAMPLIFIERS

Similarly to capacitive loading of the output, connecting a capacitor to the input of the feedback amplifier adds phase shift in the open-loop gain which in turn can lead to instability. One example of a predominantly capacitive source is a photodiode.

Photodiodes are semiconductor devices where the absorption of photons from an incident beam leads to charge pair formation. A voltage signal is obtained by passing the charge through a resistor R_{load}. As shown in Fig. B.7, the photodiode can be represented as a current source in parallel with a capacitor with a value of $\sim 10 - 100$ pF, and a resistance, typically tens of MΩ. With a load resistor in the MΩ range, the photodiode acts predominantly as a capacitor. Therefore, charge flow through the load resistor is subject to the time constant $C_{\text{ph}}R_{\text{load}}$. In

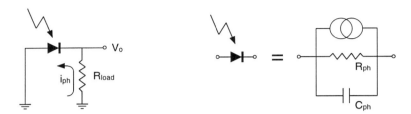

Figure B.7. **Left:** Basic photodiode circuit, generating an output voltage v_o by passing the photocurrent through a load resistor. **Right:** photodiode model.

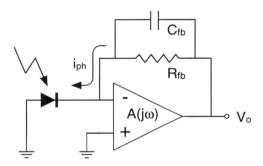

$$L(j\omega) = A(j\omega) \cdot \frac{R_{ph}}{R_{ph}+R_{fb}} \cdot \frac{1+j\omega C_{fb} R_{fb}}{1+j\omega(X_{fb}+C_{ph})R_{eq}} \; ; R_{eq} = \frac{R_{ph}R_{fb}}{R_{ph}+R_{fb}}$$

Figure B.8. **Top:** transimpedance amplifier for a photodiode in photovoltaic mode. **Bottom:** Open-loop gain for this circuit.

order to obtain a reasonably high signal level, the value of the load resistor needs to be high, which results in long time constants and inability to see transient light levels. Both high signal levels and faster response

Appendix B: Stability of Operational Amplifiers 169

can be achieved by using the arrangement shown in Fig. B.8, called a **transimpedance amplifier**. The advantage of this arrangement is in the fact that, according to Eqs. B.2,B.4 and due to the high open loop gain A of the opamp, the inverting input is at the same level as the non-inverting one, which in this case is grounded. The inverting input thus acts as a very low impedance load from the point of view of the photo-diode, which results in a very short time constant. On the other hand, due to the very large input impedance of the inverting input, practically all the photocurrent passes through the feedback resistor R_{fb}. Since one terminal of the latter is connected to the virtual ground, the voltage drop $i_{\text{ph}} R_{\text{fb}}$ will appear at the output of the amplifier.

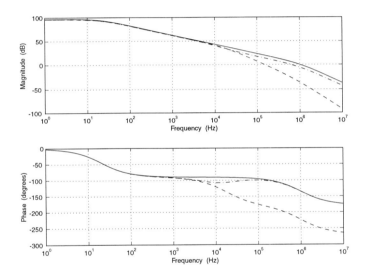

Figure B.9. Effect of capacitive input loading on the open-loop gain of the feedback amplifier, for a transimpedance amplifier used to read a photodiode. Component values used to generate these plots are: $R_{\text{ph}} = 10$ MΩ, $C_{\text{ph}} = 10$ pF, $R_{\text{fb}} = 1$ MΩ, $C_{\text{fb}} = 10$ pF. **Solid:** same L as the unloaded frequency response shown in Fig. B.6, with $|L(j\omega)| = 1$ at 1 MHz and 45° phase margin. **Long dashes:** photodiode connected, but no feedback capacitor, leading to $|L(j\omega)| = 1$ at 150 kHz and 2° phase deficit, which would cause amplifier oscillation. **Short dashes-dots:** capacitive input load compensated by adding the 10 pF feedback capacitor. This results in $|L(j\omega)| = 1$ at 600 kHz and 58° phase margin, which would restore amplifier stability.

One potential difficulty with this arrangement is its propensity for instability. In the absence of the feedback capacitor C_{fb}, the open-loop frequency response, shown in Fig. B.8, consists mainly of the open-loop gain A of the opamp and a low-pass filter with cut-off frequency $1/[2\pi R_{eq}C_{ph}]$. The 90° phase lag introduced by this low-pass filter above its cut-off frequency may erode the phase margin of the feedback amplifier to a point where instability is inevitable. This is why the feedback capacitor is necessary. Indeed, it adds a high-pass filter to the open-loop gain, which in turn results into a phase lead, restoring feedback amplifier stability. The effect of the capacitive input loading and the stabilizing effect of the feedback capacitor are illustrated in Fig. B.9.

Appendix C
Quantization Error

Most measurements carried out in today's laboratories convert raw analog quantities into digital numbers which are stored and manipulated using computers. Due to the limited number of digits available for digitization and for representing numbers in the computational environment, the values used for calculations are represented with finite resolution. The smallest difference between two numbers, related to the least significant bit, is called **quantization step**. This Appendix takes a look at the error associated with the quantization step, called **quantization error**, and at ways to reduce it.

1. ASSUMPTIONS AND NOTATIONS

1 The **signal** $x(t)$ with single-sided amplitude spectral density $X(f)$ bounded by the frequency f_x has average value

$$\bar{x} = \left[\int_0^{f_x} |X(f)|^2 \, df \right]^{\frac{1}{2}} \qquad (C.1)$$

2 The quantization step (i. e. the least significant bit) inherent to the digitization process, measured in the same units as X, is equal to b.

3 The signal x is smaller than the quantization step:

$$|x(t)| < b \qquad (C.2)$$

Note: this is a worst case assumption, because in addition to the error in estimating x, quantization introduces the highest level of distortion. This can be seen, for example, when $x = b[1/2 + 0.1 \sin \omega_o t]$. The output of the quantization process is a square wave swinging between 0 and b. If the sampling rate f_s is not much higher than ω_0/π, the higher harmonics will be aliased with the signal, making the distortion even worse.

The **resolution deficit** ρ is defined as

$$\rho \stackrel{\text{def}}{=} \frac{b}{\bar{x}} \qquad (C.3)$$

4. In addition to the signal x, the input to the quantization/sampling process contains another term $d(t)$ which will be called **dither signal**. The dither signal has single-sided spectrum $D(f)$ bounded by the frequency f_d and average value

$$\bar{d} = \left[\int_0^{f_d} |D(f)|^2 \, df \right]^{\frac{1}{2}} \qquad (C.4)$$

5. The sampling rate f_s is adequate for the dither signal as well as for the signal x, i. e. $f_s > 2\text{Max}\{f_x, f_d\}$.

6. Assuming that the true value of the measurement is uniformly distributed over the quantization step, the **quantization error** affecting each sample is

$$\delta = \frac{b}{\sqrt{12}} = \frac{\rho \bar{x}}{\sqrt{12}} \qquad (C.5)$$

7. For reasons which will become apparent later, the **dither signal** and the **sampling process** are assumed to be **uncorrelated**.

8. The dither signal is much larger than the quantization error

$$\bar{d} \gg \delta \qquad (C.6)$$

9. In order to simplify the discussion, it will be assumed that there is no noise in the system. Thus, the only variables are $x(t)$ and $d(t)$.

Appendix C: Quantization Error

Note: it sometimes happens that noise naturally present in the system performs the function ascribed here to the dither signal. However, in order to build a system that is optimal in a practical sense, it is preferable to suppress the noise below the level of $x(t)$ and design a $d(t)$ which reduces the quantization error to the desired level, as described in the remainder of this Appendix.

2. QUANTIZATION ERROR SUPPRESSION

2.1. Signal/Error Ratio with Dithering and Filtering

In the absence of noise, the average error in the estimate of $x(t)$ generated by the quantization/sampling process is:

$$\bar{\epsilon} = \sqrt{\bar{d}^2 + \delta^2} \stackrel{\text{Eq. C.6}}{\approx} \bar{d}\left(1 + \frac{1}{2}\frac{\delta^2}{\bar{d}^2}\right) \approx \bar{d} + \left[\frac{1}{2}\frac{\delta}{\bar{d}}\right]\delta \stackrel{\text{def}}{=} \bar{d} + \epsilon_q \quad \text{(C.7)}$$

where $\epsilon_q = \delta^2/2\bar{d}$ is the residual quantization error. Eq. C.7 shows that the additive contribution of quantization to the error in estimating $x(t)$ is reduced by $\delta/2\bar{d}$ in the presence of the dither signal $d(t)$. Thus, if the dither-to-quantization error ratio is large enough (Assumption 8), the error affecting the estimate of $x(t)$ is dominated by the dither component, while the quantization error has been substantially attenuated.

If digitization/sampling is followed by a filtering process, the resulting signal-to-error ratio R is

$$R = \frac{\bar{x}}{\bar{\epsilon}} = \frac{\left[\int_0^{f_x} |\mathcal{F}(j\omega)X(f)|^2\,df\right]^{\frac{1}{2}}}{\left[\int_0^{f_d} |\mathcal{F}(j\omega)D(f)|^2\,df\right]^{\frac{1}{2}}} \quad \text{(C.8)}$$

where $\mathcal{F}(j\omega)$ is the frequency response of the filter.

Eq. C.8 shows that if $X(f)$ and $D(f)$ differ significantly from zero over different non-overlapping frequency ranges, the filter response $\mathcal{F}(j\omega)$ can be chosen such as to pass $x(t)$ and reject $d(t)$, and thus lead to a sizable signal-to-error ratio R.

Note: In order to suppress the quantization error, Eq. C.7 has to hold; this, in turn, is predicated upon the assumption that the quantization error and the dither signal are uncorrelated. This requires that the dither signal and the sampling process be uncorrelated (Assumption 7).

2.2. Signal Recovery

As shown by Eq. C.7, the effect of quantization on the output is reduced when a large dither signal is present at the input of the digitizer. Furthermore, according to Eq. C.8, the dither signal can be suppressed by post-digitization filtering, if the dither signal $d(t)$ and the useful signal $x(t)$ have non-overlapping spectra. In order to recover the signal $x(t)$, the sum of the two terms in Eq. C.7 needs to be reduced to a level where the signal-to-noise R reaches a predetermined level consistent with the performance requirement for the measurement. In order to make the discussion concrete, it will be assumed that the two contributions to the residual error will be made equal to $\bar{x}/2R$. Of course, any ratio between the two terms is acceptable, as long as the sum is small enough to ensure the prescribed signal/error ratio R.

Signal recovery can be conducted as follows:

1. Given a quantization resolution deficit ρ and a required signal-to-error ratio R, determine the dither amplitude $\bar{d}$ which suppresses the quantization error down to a level corresponding to $\epsilon_q = \bar{x}/2R$. Using Eqs. C.3,C.5,C.7:

$$\bar{d} = \frac{R\rho b}{12} \quad \text{(C.9)}$$

2. Determine the conditions under which $\bar{d}$ can be suppressed to a value $\bar{d}_{residual} = \bar{x}/2R$. This step is divided into five sub-steps:

   (a) Determine the required attenuation factor A. From Eqs. C.3,C.9

$$A = \frac{R^2 \rho^2}{6} \quad \text{(C.10)}$$

   (b) Determine the frequency response of the steepest low-pass filter which can attenuate $d(t)$ by the required factor A, while passing the signal $x(t)$ with little or no attenuation, and which does not violate other constraints imposed on the measurement. Examples of such constraints are:

Appendix C: Quantization Error 175

- Signal recovery has to be carried out in real time, and the delay caused by the filtering cannot exceed a certain value. This constraint imposes a limitation on how steep the filter can be.

- Signal recovery has to be carried out in real time, and both the delay and the instantaneous phase shift introduced by the filter cannot exceed certain values, because the signal $x(t)$ is to be used in a feedback control loop. This constraint can result in an even more severe limitation to the steepness of the filter.

(c) Taking into account the frequency response of the filter, determine the lowest frequency to which the spectrum of $d(t)$ can extend, while the required attenuation factor A can still be achieved.

(d) Specify the spectrum of $d(t)$ as band-limited random noise, with the lowest frequency as specified at the previous point, and with average amplitude specified by Eq. C.9.

(e) Specify the minimum sampling rate capable to provide a reasonable representation of $d(t)$, given its spectrum as specified above.

The main ingredients of the above procedure for reducing the quantization error are:

- **Dithering**, whereas the quantization error is essentially suppressed and replaced with a much larger dither signal. The dither signal is preferably colored random noise. Its spectrum $D(f)$ is chosen so that it does not overlap with the spectrum of the variable of interest, $X(f)$.

- Post-digitization **filtering** is used to suppress the dither signal $d(t)$, while preserving the "useful" signal $x(t)$.

- **Oversampling**. As mentioned above, the sampling rate has to ensure that neither $x(t)$ nor $d(t)$ are subject to aliasing in the digitization process, i.e. $f_s > 2\text{Max}\{f_x, f_d\}$. If adequate suppression of $d(t)$ by post-digitization filtering requires that $f_d \gg f_x$, the sampling rate has to be much higher than $2f_x$. The signal $x(t)$ is then said to be **oversampled**.

176   FEEDBACK CONTROL SYSTEMS

3. EXAMPLE

Consider an application where changes of the optical path in a system need to be controlled to 1 nm, with a control bandwidth of 1 kHz, and known to 10 pm over time scales of 1000 seconds.[1] The sensor is an interferometer with 10 kHz sampling rate and read-out resolution of 500 pm. Using the language of the previous sections, the terms of this problem are:

- $b = 500$ pm
- $\bar{x} = 10$ pm
- $f_x = 0.001$ Hz
- $f_s = 10$ kHz

From Eq. C.3, the resolution deficit in this problem is $\rho = 50$. For unit signal-to-error $R = 1$, the required dither amplitude, calculated from Eq. C.9, is:

$$\bar{d} = 2083 \text{ pm} = 2.083 \text{ nm} \quad (C.11)$$

The dither should occur outside the control band of 1000 Hz, otherwise the loop would suppress the dither to some extent, which would prevent reaching the desired level of quantization error suppression. The spectrum of the dither signal can be placed, for example, in a 1 kHz band centered at 5 kHz, where it would be appropriately digitized given the sampling rate of 10 kHz. The spectral amplitude of the dither signal within that band would be:

$$D(f) = 2083 \text{ pm}/\sqrt{1000 \text{ Hz}} = 65.9 \text{ pm}/\sqrt{\text{Hz}} \ , \quad 4500 \text{ Hz} \leq f \leq 5500 \text{ Hz} \quad (C.12)$$

In order to obtain knowledge of the optical path difference to 10 pm, the interferometer readout needs to be filtered outside the control loop, possibly by post-processing, for suppression of the dither amplitude. The required suppression factor, calculated by using Eq. C.10, is:

$$A = 417 \quad (C.13)$$

This level of attenuation should be easy to achieve, given the large frequency separation, 6 decades, between the center of the dither spectrum

---

[1] 1 nm = 10^{-9} m, 1 pm = 10^{-12} m.

Appendix C: Quantization Error 177

and the frequency where the resolution needs to be 10 pm. However, the 2.083 nm dither amplitude conflicts with the requirement of controlling the optical path to 1 nm. Increasing the resolution of the digitizer to 250 pm (one additional bit) would remove this conflict. Indeed, according to Eq. C.9, $\bar{d} \propto b^2$, since $\rho = b/\bar{x}$. Therefore, if the resolution is improved to 250 pm, the dither amplitude can be reduced to 0.521 nm, which would be acceptable.

Index

actuator, 4
    collocation with sensor, 56
    desaturation, 104, 109, 116, 117
    desaturation ratio, 116
    driver, 9–11, 52, 72, 73, 81, 88–91, 94, 106
    fast/slow combination, 68
    for changing the laser frequency, 45
    frequency range overlap, 117
    insufficient range, 83, 87, 89
    range, 15, 57, 67, 95, 96
    saturation, 15, 108
    speed, 67
aliasing, 121
amplifier saturation, 106

Bode plot, 21, 44
    asymptotic, definition, 24
    asymptotic, example, 25
    example, 21
Bode step, 70, 71
by-pass arrangement, 107

calibration, 138
characteristic equation, 18
characteristic polynomial, 18
closed loop, 4, 9
    failure of, *see* system integration
    simplified diagrams, 14
    stability, 33, 68
    transfer function, 11
command, 4
compensator, 4, 10
    as filter, 10
    digital, 119
    digital transfer function implementation, 123
    digital, components of, 120
    digital, delay due to sampling, 124
    digital, reasons for use, 122
    gain adjustment, 73
    manual gain control, 84, 86
    need for redesign, 61
    noise, impressed on output, 12
    saturation of, 61
    saturation, upon lock acquisition, 75
    simplification, by plant design, 56
    simplification, by sensor/actuator collocation, 56
    specification, 72
    test input, 73
control hand-off, 104
control input, 4, 11–14
control system architecture, 56
control system design
    brief description, 51
    broad definition, 54
    input information, 57
    iteration, 57
    main topics, 52
    outline, 59
    priorities in the design, 51
    schematic representation, 55
    steps, 56
controller, 4
convolution product, 5
corner frequency, 25, 115
correction signal, 4, 10, 12
    equation for, 12

dB, definition, 21
diagram
    addition, 7
    multiplication, 7

of general feedback loop, 5–7, 10, 11
of simplified closed loop, 14
dispersion relations, 22, 157
disturbance
    measurement of, 91
    suppression, 4, 12, 14, 15, 24, 26, 51, 81, 99, 100, 143
    suppression by loop gain, 12
    suppression, arrangement, 41

error
    drift, 5–7
    noise, 5–7
    of reference, 62
    of sensor electronics, 62
    offset, 5–7
    tracking, *see* tracking, acuracy
error signal, 4, 10
example
    laser stabilization, 65, 66

feedback control, 4
feedback loop stability, 17
forcing function, 5, 18, 26, 156
    Laplace transform of, 6
    spectrum of, 26
frequency
    Fourier, 44, 46, 47, 67
    in-band, 48
    out-of-band, 48
    stability, 47
frequency response
    definition, 6
    open-loop, 14
    piecewise measurement, 86

gain, *see also* open loop
    increase at low frequency, 87
gain margin, 20–22

impulse response, 5
instability, 4, 51
    at hand-off, 104, 143
    causing closed-loop failure, 83
    causing oscillation build-up, 84
    due to excessive phase lag, 83
    due to feedback, 27
    due to flexible elements, 147, 151
    due to plant resonances, 56
    due to positive feedback, 83
    due to resonances, 102
    due to resonances, prevention of, 152
    due to very high phase lag, 32
    for gain too high or too low, 84
    intermittent, 75

opamp with capacitive load, 166, 167
opamp with capacitive source, 167, 169

lag-lead circuit, 87
laser noise suppresion, example, 63
lock acquisition, 36
    and gain boost switch, 108
    and system architecture, 54
    arrangement for inducing, 77
    as part of system architecture, 56
    difficulty, 95
    efficiency, means of improvement, 96
    induced, algorithm, 76
    need for, 66
    spontaneous, 76

MIMO, 99, 141
momentum compensation, 151, 152

network analyzer, 79, 90, 131
noise
    design considerations, 62
    dominant contributions, 54
nonlinearity
    associated with power supply voltage, 123
    associated with quantization step, 123
Nyquist criterion, 17, 34, 151
Nyquist criterion using Bode plots, 22

open loop
    frequency response, measurement of, 57
    gain, adjustments to, 96
    gain, example of lower bound derivation, 68
    gain, incorrect value, 84
    gain, minimum magnitude, 68
    gain, specification criteria, 68
    transfer function, 11
    transfer function measurement, 90

phase margin, 20, 22–24, 70, 71, 74, 117, 150
    "optimum value", 70
    and time-domain behavior, 23
    at PID hand-offs, 116
    decrease by series gain boost stage, 106
    decrease with increasing overall gain, 86

Index

degradation by amplifiers, 75
degradation by capacitive source, 169
degradation by digital compensator, 122
degradation by propagation delay, 123
erosion, 66
excessive, 71
increase, to reduce ringing, 96
minimum values, 24
of opamp, 164
opamp with capacitive load, 165
restoration, using buffer resistor, 166
phase reserve, 123
plant, 3
prefilter, 71
procedure
    for digital compensator specification, 119

quantization
    error, 171
    step, 171

range
    of actuator, 67
    of environmental parameters, 58
    of sensor, 66
reactuation, 151, 152
robustness, 21
    and margins, 22, 24
    of tracking, 94
    with digital compensators, 123

sampling rate, 120, 124
    and D/A ringing, 132
    and resolution, 126
    and small signal distortion, 172
    specifying of, 133
saturation
    of amplifier, 106
    recovery from, 61
search algorithm, 36
sensor, 3

collocation with actuator, 56
error, 10, 44
error, from tracking requirement, 62
gain, 44, 64
head, 64
multi-stage design, 64
noise, 10, 44
noise influencing tracking estimate, 93
noise, attenuation of, 13
noise, impressed on output, 12
nonlinearity, examples, 66
outside the loop, 93
range, 36, 44
saturation, upon lock acquisition, 75
transfer function, 64
stability
    condition, 18, 19
    definition, 18, 27
    described with differential equations, 17
    Nyquist criterion, 19
system architecture, 54, 56
system integration
    cause of closed loop failure, 83
    insufficient gain, 86
    necessary equipment, 78
    troubleshooting procedure, 82

tracking
    accuracy, 12
    of aircraft, example, 35
    robustness, 94
    the control input, 12
    the reference, 15
transfer function, 6
    closed-loop, 11, 15
        for disturbance, 15
        for reference input, 15
    of FCS components, 11
    of series and parallel blocks, 7
    open-loop, 11, 12, 14

uncertainty
    of disturbance, 58